# DICTIONARY
## OF
# MATHEMATICAL GAMES, PUZZLES, AND AMUSEMENTS

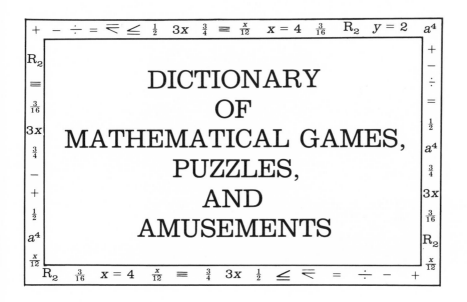

# DICTIONARY
## OF
# MATHEMATICAL GAMES,
# PUZZLES,
## AND
# AMUSEMENTS

HARRY EDWIN EISS

GREENWOOD PRESS

New York • Westport, Connecticut • London

**Library of Congress Cataloging-in-Publication Data**

Eiss, Harry Edwin.
  Dictionary of mathematical games, puzzles, and
amusements.

  Bibliography: p.
  Includes index.
  1. Mathematical recreations—Dictionaries.
  2. Puzzles—Dictionaries.   I. Title.
  QA95.E37     1988        793.7′4′0321        87-280
  ISBN 0-313-24714-5 (lib. bdg. : alk. paper)

British Library Cataloguing in Publication Data is available.

Library of Congress Catalog Card Number: 87-280
ISBN: 0-313-24714-5

First published in 1988

Greenwood Press, Inc.
88 Post Road West, Westport, Connecticut 06881

Printed in the United States of America

The paper used in this book complies with the
Permanent Paper Standard issued by the National
Information Standards Organization (Z39.48-1984).

10 9 8 7 6 5 4 3 2 1

## Copyright Acknowledgment

Illustrations appearing on pages 175 and 178 are taken from Lewis Carroll, *Through
the Looking-Glass: And What Alice Found There,* 1896. Reprinted in *The Complete
Works of Lewis Carroll,* Random House, n.d.

To my wife, Betty

# CONTENTS

# PREFACE

Zeno of Elea gazes out over the beautiful country of ancient Greece. He has recently defended the philosophy of his teacher Parmenides against the views of Pythagoras, Heraclitus, and Anaxagoras. Perhaps Parmenides' view of the world as a solid, finite, uniform entity, without time or motion, has not answered the questions of all existence once and for all, Zeno thinks, but surely it makes more sense than the other philosophies.

Motion must be only an illusion. An arrow at each instant in its path must occupy a specific place in space. It is, therefore, always at rest. Puzzling, though, Zeno speculates. And what if space is divided up into infinitely smaller particles? Then a man could never walk across the street, because before he could get all the way across he would have to get half way across, and before he could get half way across he would have to get half of half way across, and on and on into infinity.

Simple paradoxes? Easily answered? Well, it took some 2,400 years for someone to arrive at an answer. In 1873, Georg Cantor came up with the Continuum Hypothesis, which is based on the theory that, when dealing with the infinite, it is necessary to replace some of the rules used for finite mathematics. One of the most important is that an infinite collection can be put into a one-to-one correspondence with a part of itself. In other words, it is possible for there to be an infinite number of integers, and it is also possible for there to be an infinite number of even integers. Thus, there are infinities beyond infinities. The first infinite set Cantor designated as aleph-null, the first infinite set beyond aleph-null as aleph-one.

This is not the place to get into a highly sophisticated discussion of Cantor's theories (refer to the entry on Paradoxes of the Infinite for a more detailed overview). The point is that it took millennia for someone to come up with a solution to Zeno's paradoxes (refer to the entry on Zeno's Paradoxes for further discussion of them), and that Cantor's theories are no more universally accepted by contemporary mathematicians than Zeno's were by his contemporaries.

David Hilbert, the brilliant leader of what is called the Axiomatic Group of twentieth-century mathematicians, praises Cantor: "No one shall expel us from the paradise which Cantor has created for us."[1] However, equally brilliant mathematicians condemn Cantor's theories. Henri Poincaré has stated, "Later generations will regard [Cantor's] *Mengenlehre* as a disease from which one has recovered."[2] Kurt Gödel, who proved that Cantor's assumption that there is no aleph between aleph-null and the power of the continuum could be assumed to be true, and Paul J. Cohen, who proved the opposite, both believe that Cantor's Continuum Hypothesis will ultimately be proven false.

The questions remain, and they rest at the very foundations of mathematics. But what does all this have to do with mathematical games, puzzles, and amusements? First of all, Zeno's paradoxes are among the most commonly included subject matter in recreational mathematics, and Cantor's theories serve as the basis for numerous problems included in recreational mathematics. Furthermore, since the first third of the twentieth century, there has been a shift in recreational mathematics from less sophisticated amusements in number curiosities, mazes, simple geometric puzzles, arithmatic story problems, card tricks, magic squares, trisection of the angle, squaring the circle, duplication of the cube, Pythagorean triples, board games, and so on, to at times highly sophisticated explorations of number theory, game theory, graph theory, topological problems, flexagons, logical paradoxes, fallacies of logic, paradoxes of the infinite, and so on.

Mathematical games, puzzles, and amusements go back to the beginning of recorded history, many of them, such as Zeno's paradoxes, reappearing time and time again throughout history. Little is known about the mathematical recreations in existence during the so-called Dark Ages, but by the Middle Ages, notable names, such as Fibonacci (Leonardo of Pisa; 1202), who is famous for the Fibonacci series, and Rabbiben Ezra (1140), appear. Gerolamo Cardan (1501–1576) and Nicholas Tartaglia (Niccolo Fontana; 1500–1557) took part in a mathematical duel as the result of Cardan stealing Tartaglia's solution to a cubic equation.

In the seventeenth century, books began appearing devoted solely to recreational mathematics. Claude Gaspard Bachet de Méziriac (1581–1638) wrote *Problèmes plaisans et délectables qui se font par les nombres* (1612), which went through five editions, the most recent in 1959, and served as a source for similar collections. It included card tricks, watch-dial puzzles, weight problems, and difficult crossings.

Van Etten (Jean Leurechon) wrote *Récréations mathématiques* in 1624, based mainly on Bachet. It went through thirty editions before 1700. Claude Mydorge wrote *Examen du livre des récréations mathématiques* (1630), and Denis Henrion wrote *Les Récréations mathématiques avec l'examen de ses problèmes en arithmétique, géométrie, méchanique, cosmographie . . .* (1659)—both based on van Etten. In Germany, Daniel Schwenter put together a collection of recreational problems based on Leurechon's book. These were published after his death as *Deliciae Physico-mathematicae oder Mathematische und Philosophische Erquickstunden* (1636) and were extremely popular. Subsequently, an enlarged edition was published.

Mario Bettini published *Apiaria Universae Philosophiae Mathematicae in Quibus Paradoxa et Nova Pleraque Machinamenta Exhibentur,* a two-volume set, in 1641–1642, and followed this with a third volume, *Recreationum Mathematicarum Apiaria Novissima Duodecim . . .* in 1660.

In England, William Leybourn published *Pleasure with Profit: Consisting of Recreations of Divers Kinds, viz. Numerical, Geometrical, Mechanical, Statical, Astronomical, Horometrical, Cryptographical, Magnetical, Automatical, Chymical, and Historical* in 1694.

In the eighteenth century, England began publishing mathematics recreations in earnest. Edward Hatton, Thomas Gent, Samuel Clark, William Hooper, and Charles Hutton are some of the more important authors. The most important work of the time, however, was that of Jacques Ozanam, *Récréations mathématiques et physiques* (1694), a four-volume set that went through numerous editions, eventually being translated into English by Charles Hutton (1803, 1814) and again by Edward Riddle (1840, 1844).

The second half of the nineteenth century produced a number of important writers in recreational mathematics, the most famous being Lewis Carroll (C. L. Dodgson). But others, notably Edouard Lucas, author of *Récréations mathématiques* (1882–1894), a four-volume set, were also producing classic works in the field. By the turn of the century, Sam Loyd and his son, Sam Loyd, Jr., were publishing hugely successful volumes of puzzles and holding a friendly rivalry with Henry Ernest Dudeney, a British collector of games and puzzles. These creators and collectors were the forerunners of contemporary recreational mathematics. Others, however, were also important. W. W. Rouse Ball published *Mathematical Recreations and Essays* in 1892, using a more scholarly approach; it became a classic and is still a standard reference. Maurice Kraitchik, editor of *Sphinx,* published *Mathematical Recreations* (1942), which has since been revised and remains a standard reference. Fred Schuh published *Wonderlijke Problemen; Leerzaam Tijdverdrijf Door Puzzle en Spel* (1943), another important text on mathematical recreations, translated and republished in English by Dover Publications in 1968, once again, a more esoteric approach.

These books, and others, reveal the shift to more sophisticated recreations. H. Steinhaus' *Mathematical Snapshots* (1950), Joseph S. Madachy's *Mathematics on Vacation* (1966), and the numerous books by Martin Gardner from the 1950s to the present all attempted to deal with highly intellectual mathematical concepts in an attractive, nonspecialist manner.

This brief history would not be complete without mentioning the excellent, continually updated bibliography of recreational mathematics put out by W. L. Schaaf, an indication of the popularity of the field today, as are the numerous magazines containing articles on recreational mathematics, for example, *American Mathematical Monthly, Arithmetic Teacher, Fibonacci Quarterly, Journal of Recreational Mathematics, Mathematics Magazine, Recreational Mathematics Magazine, Scientific American,* and *Scripta Mathematica.*

Game theory has recently become an important branch of mathematics. It attempts to analyze problems of conflict by abstracting common quantifiable strategic features. These features can then be reformulated in mathematical language as games, since they are patterned on such real games as bridge and poker. The French mathematician Émile Borel first put forth a theory of games of strategy in 1921. However, it was John von Neumann who independently derived the cornerstone for the theory and presented it in 1928. This cornerstone is what is referred to as the Min-Max theorem, i.e., each player attempts to minimize the opponent's maximum gain. It is not necessary to go into the complexities of the theory here, but rather simply to point out that game theory opens up the door to a mathematical analysis of such wide-ranging concerns as sociology, psychology, politics, and war—certainly matters of serious concern, and all tied together with recreational mathematics—which leads me to the following passage written by Henri Poincaré (directed to science, but also applicable to mathematics and even to recreational mathematics):

Science keeps us in constant relation with something which is greater than ourselves; it offers us a spectacle which is constantly renewing itself and growing always more vast. Behind the great vision it affords us, it leads us to guess at something greater still; this spectacle is a joy to us, but it is a joy in which we forget ourselves and thus it is morally sound.

He who has tasted of this, who has seen, if only from afar, the splendid harmony of the natural laws will be better disposed than another to pay little attention to his petty, egoistic interests. He will have an ideal which he will value more than himself, and that is the only ground on which we can build an ethics. He will work for this ideal without sparing himself and without expecting any of those vulgar rewards which are everything to some persons; and when he has assumed the habit of disinterestedness, this habit will follow him everywhere; his entire life will remain as if flavored with it.

It is the love of truth even more than passion which inspires him. And is not such a love an entire code of morality? Is there anything which is more important than to combat lies because they are one of the most common vices in primitive man and one of the most degrading? Well! When we have acquired the habit of scientific methods, of their scrupulous exactitude, of the horror of all attempts to deflect the course of experiment; when we have become accustomed to dread as the height of ignominy the censure of having, even innocently, slightly tampered with our results; when this has become with us an indelible professional habit, a second nature: shall we not then reveal in all our actions this concern for absolute sincerity to the extent of no longer understanding what makes other men lie? And is this not the best means of acquiring the rarest, the most difficult of all sincerities, the one which consists in not deceiving oneself?[3]

These lofty realms truly are a part of recreational mathematics, not just in the paradoxes of Zeno, but in the joy to be found in Pascal's triangle or the discovery of a new knight's tour or the successful squaring of the square or the recent proof of the four-color map problem or the development of the Franklin squares or the discovery of an additional prime number or, perhaps, even a resolution to Cantor's Theory of the Continuum, leading to a better understanding of the problems Zeno contemplated as he gazed over the beautiful country of ancient Greece.

This book does not offer a solution to Zeno's paradoxes or, for that matter, even a new knight's tour. Rather, it contains a collection of the various mathematical games, puzzles, and amusements that have led people to such new solutions. The goal has been to bring together the creations of others. Certainly, not all of the mathematical recreations of all time are included, nor are all of the contemporary mathematical recreations included. That would be an impossible task. The attempt, rather, has been to collect a good number of recreations of various types. A generous amount of discussion is given to such classical problems as the trisection of an angle and Euclid's Theory of Parallels, and to such important games as chess and nim. Less esoteric recreations, such as checkers and dominoes, also receive some attention. Topological concerns, such as the Möbius strip, are included, as well as such arithmetic curiosities as narcissistic numbers. The origins and history of each recreation are included if known.

Similar recreations have been cross-referenced to enable the reader to easily refer to other puzzles, games, and amusements of the same type. A bibliography follows each entry to direct the reader to further sources of information.

I thank Marilyn Brownstein, Acquisitions Editor at Greenwood Press, for seeing the value of the book and encouraging her company to publish it, and Michelle Scott, Senior Production Editor at Greenwood Press, for seeing the work through to its final form. The librarians at Northern Montana College under the direction of Terrence A. Thompson, in particular Vicki Gist, deserve a great deal of thanks for their efforts in obtaining the research materials I needed. My wife and children have been forced to accept my lack of time for them, as well as a kitchen table piled high with books and paper and typewriter, while I completed the project, and I want them to know I appreciate their understanding and encouragement. Finally, I want to thank the many brilliant mathematicians I have come to know through the years for the entertainment and knowledge they have given me.

**NOTES**

1. Morris Kline, *Mathematics in Western Culture* (New York: Oxford University Press, 1953), p. 397.

2. Ibid.

3. Henri Poincaré, *Mathematics and Science: Last Essays,* trans. by John W. Bolduc (orig. publ. 1913; New York: Dover, 1963), p. 105.

# DICTIONARY
## OF
# MATHEMATICAL GAMES,
## PUZZLES,
## AND
# AMUSEMENTS

# A

**ABC WORDS,** an acronym for alphabetically balanced combination words, are words whose average numerical value is $13\frac{1}{2}$.

The letters of the alphabet are sequentially assigned numerical values (i.e., A = 1, B = 2, C = 3, and so on). Numbers are then substituted for the letters of a word. If the total of the sum of all the numbers of a word divided by the number of letters in the word equals $13\frac{1}{2}$, then the word is an ABC word.

Dmitri Borgmann, the creator of the activity, in *Beyond Language*, offers the suitably appropriate word "love" as an example of a perfectly balanced word: L = 12, O = 15, V = 22, E = 5. All of the letters added together equal 54, which divided by 4 equals $13\frac{1}{2}$.

Borgmann suggests a number of mathematical balances that may be included in the equation. The rules may include the need for an equal number of letters to come from each half of the alphabet. It may even be established that the letters for the first half of the word come from the first half of the alphabet and the letters for the second half of the word come from the second half of the alphabet, e.g., "bins." It may be established that the letters perfectly balance one another, e.g., "Zola." It may be established that the words have no balanced letters, e.g., "rope." It may be established that the words use no letter more than once, e.g., "smog." It may be established that the words have an equal number of vowels and consonants, e.g., "tone."

For similar mathematical word play refer to ACE Words; Centurion; Numwords. For mathematical mysticism involving letters and words refer to Numerology.

BIBLIOGRAPHY

Borgmann, Dmitri. *Beyond Language: Adventures in Word and Thought*. New York: Charles Scribner's Sons, 1967.

Brooke, Maxey. *150 Puzzles in Crypt-Arithmetic*. 2d rev. ed. New York: Dover, 1963.
McKechnie, Jean L., et al., eds. *Webster's New Twentieth Century Dictionary of the English Language: Unabridged*. 2d ed. New York: World Publishing Co., 1971.
Stein, Jess, ed. *The Random House Dictionary of the English Language*. New York: Random House, 1983.

**ACE WORDS,** an acronym for alphabetically constant entity words, involve a form of mathematical word play.

The object is to stabilize a word of seven or more letters mathematically by achieving a constant difference in the numbers used to designate the letters of the word. In each case, the second numeral, indicating the number of digits in the first constant difference level, must be more than one-half of the first numeral, which indicates the number of letters in the original word.

It works as follows: The letters of the alphabet are assigned numeral equivalents (i.e., A = 1, B = 2, C = 3, and so on). The appropriate numerals are then substituted for the letters of a word of at least seven letters. Dmitri Borgmann, who thought up the activity, uses "inkling" as an example of an ACE 7-5 word:

INKLING

$I = 9, N = 14, K = 11, L = 12, I = 9, N = 14, G = 7$

After the letters have been assigned their corresponding numerals, the difference between successive letters is determined mathematically by subtracting the smallest from the largest:

$14 - 9 = 5, 14 - 11 = 3, 12 - 11 = 1, 12 - 9 = 3, 14 - 9 = 5, 14 - 7 = 7$

This produces a 5, 3, 1, 3, 5, 7 sequence of differences at the 7-6 level. By repeating the process the following constant difference is achieved at the 7-5 level:

$5 - 3 = 2, 3 - 1 = 2, 3 - 1 = 2, 5 - 3 = 2, 7 - 5 = 2$

(2, 2, 2, 2, 2)

Borgmann divides ACE words into three orders of reduplication. In the first order, exact reduplication occurs (e.g., "ring-ring"). In the second order, there are three subdivisions of words which have only one letter changed. The first of these contains all of the words where the reduplication is done on purpose (e.g., "flip-flop"). The second contains all of the words where the reduplication is accidental but occurs at the exact center (e.g., "shoeshop"). The third consists of words where the accidental reduplication does not fall along well-defined syllable or word divisions (e.g., "jingling"). The third large order of reduplication contains the words that have more than one letter changed (e.g., "postpone").

Borgmann notes that there are ACE words that contain no reduplication (e.g., "America").

For similar mathematical word play refer to ABC Words; Centurion; Numwords. For mathematical mysticism involving letters and words refer to Numerology.

BIBLIOGRAPHY

Borgmann, Dmitri A. *Beyond Language: Adventures in Word and Thought.* Charles Scribner's Sons: New York, 1967.
McKechnie, Jean L., et al., eds. *Webster's New Twentieth Century Dictionary of the English Language: Unabridged.* 2d ed. New York: World Publishing Co., 1971.
Stein, Jess, ed. *The Random House Dictionary of the English Language.* New York: Random House, 1983.

**ACHI** is a board game.

According to Gyles Brandreth it is popular among Ghanaian school children. Two players start with four counters each and sit at opposite sides of the following board:

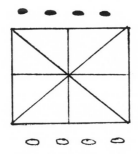

The players move alternately. They begin by placing their four counters on eight of the nine points of the board. The object is to get three markers in a row. This is done by moving one marker at a time along a line to a free point.

For similar recreations refer to Geometric Problems and Puzzles; Topology.

BIBLIOGRAPHY

Brandreth, Gyles. *Indoor Games.* London: Hodder & Stoughton, 1977.
_____. *The World's Best Indoor Games.* New York: Pantheon Books, 1981.

**ACHILLES AND THE TORTOISE.** See ZENO'S PARADOXES.

**AGE PUZZLES.** See ARITHMETIC AND ALGEBRAIC PROBLEMS AND PUZZLES.

**ALBERTI DISK.** See CRYPTARITHMS, CRYPTOGRAPHY, CONCEAL-MENT CIPHERS, SUBSTITUTION CIPHERS, TRANSPOSITION CIPHERS, AND CODE MACHINES.

**ALGEBRAIC FALLACY.** See FALLACIES.

**ALL CRETANS ARE LIARS PARADOX.** See LOGICAL PARADOXES.

**ALL FIVES.** See DOMINOES.

**ALPHAMETIC.** See CRYPTARITHMS, CRYPTOGRAPHY, CONCEAL-MENT CIPHERS, SUBSTITUTION CIPHERS, TRANSPOSITION CIPHERS, AND CODE MACHINES.

**AMICABLE NUMBERS.** See NUMBER PATTERNS, TRICKS, AND CURIOSITIES.

**ANTIMAGIC SQUARES.** See MAGIC SQUARES.

**ARITHMETICAL PROGRESSIONS.** See NUMBER PATTERNS, TRICKS, AND CURIOSITIES.

**ARITHMETIC AND ALGEBRAIC PROBLEMS AND PUZZLES** are those recreations involving integers and the operations of addition, subtraction, multiplication, and division, and general statements of relations, using letters and other symbols to stand for specific numbers, values, and so on.

The two following *clock puzzles* appeared in *The Mathematical Puzzles of Sam Loyd:* If both hands of a clock are at the same distance from the six hour mark at almost twenty minutes after eight, exactly what time is it? The solution: $18\frac{6}{13}$ minutes past eight. If one watch goes two minutes per hour too slow, and the other goes one minute per hour too fast, how long will it take for them to be exactly one hour apart? The solution: three minutes per hour, so twenty hours.

The following problem comes from Henry Ernest Dudeney, *Amusements in Mathematics:* A clock hangs on a 71', 9" by 10', 4" wall. The hands of the clock are pointing in opposite directions and are parallel to one of the diagonals of the wall. What time is it? The solution: $43\frac{7}{11}$ minutes past 2:00.

Here is another one from Sam Loyd. If the two hands of a clock are together at exactly 12:00, when will the hands next be together again? Solution: 5 minutes, $27\frac{3}{11}$ seconds past 1:00, i.e., the minute hand travels twelve times faster than the hour hand.

The following problem from Sam Loyd is a sample of what are often called *money problems:* A man gives the postmaster a dollar bill and asks for some two-cent stamps, ten times as many one-cent stamps, and the rest in five-cent stamps. Solution: five two-cent stamps, fifty one-cent stamps, and eight five-cent stamps.

The following comes from Martin Gardner, *Mathematical Magic Show,* who credits its creation to Ronald A. Wohl and Carl J. Coe: Two brothers inherit a herd of sheep. They sell all of the sheep, receiving for each sheep the same number of dollars as there were sheep in the herd. The money is paid to them in $10 bills, except for an excess amount, which is paid in silver dollars. They divide the bills, but one $10 bill plus the silver dollars is left over. To even things up, one brother takes the $10 bill, gives the other the silver dollars, and writes out a check to even out the difference. What is the value of the check?

The solution: $n$ = the number of sheep; $n^2$ = the total number of dollars received; the total amount must contain an odd number of $10 bills; thus, the number of sheep must end with a digit the square of which contains an *odd* number of 10s (only 4 and 6 satisfy this; 16, 36); since both squares end in 6, the excess has to be $6; thus, the check has to equal $2.

Martin Gardner includes the following in *Mathematical Puzzles:* A customer asks for change for a dollar. The cashier says she doesn't have the correct change. The customer asks for change for a half dollar. The cashier says she cannot change it. In fact, she says that she cannot make change for a quarter, dime, or nickel. Yet she says she has $1.15 worth of change. What coins are in the register? Solution: one half-dollar, one quarter, and four dimes.

The following *probability puzzle* is from *Mathematical Puzzles:* If three coins are tossed into the air and allowed to fall as they may, will it be a good bet to bet two to one that they will all fall heads or tails (will not be mixed)? Solution: There are eight possible ways for the coins to fall (one all heads, one all tails, and six mixed). Thus, in two out of eight or one out of four times they will all be identical. It would be a bad bet.

The following example comes from Sam Loyd: At the end of a party, six hats are left in the cloak room, but the men they belong to are so drunk that they cannot determine whose hat belongs to whom. As it turns out, no one gets his own hat. What are the chances of this happening? Solution: 265/720; $n$ = the total number of ways hats can be selected; $n! \left( 1 - \dfrac{1}{1!} + \dfrac{1}{2!} + \dfrac{1}{3!} \ldots \pm \dfrac{1}{n} \right)$.

Martin Gardner includes this standard probability puzzle in *Mathematical Puzzles:* If a die is rolled nine times and comes up 1 each time, what is the likelihood it will come up 1 the tenth time? Solution: $\frac{1}{6}$; the probability remains the same regardless of the number of rolls.

Clark Kinnard includes the following *time problem* in *Encyclopedia of Puzzles and Pastimes:* Three boys fill a 53-gallon tank. The first boy brings a pint every three minutes; the second a quart every five minutes; the third a gallon every seven minutes. How long does it take, and who pours the final amount? Solution: 230 minutes; the second boy adds the final quart.

Martin Gardner includes the following in *Mathematical Puzzles:* If you have three cats that can catch three mice in three minutes, how many cats will you need to catch 100 mice in 100 minutes?

Solution: The question is ambiguous. If the three cats catch one mouse every minute, then they can catch 100 mice in 100 minutes. However, if each cat takes three minutes to catch a mouse, then they will have caught 99 mice in 99 minutes. Then they still have one mouse to catch—how long will it take? Three minutes? One minute?

The following is a traditional *age puzzle* included in James F. Fixx, *More Games for the Super-Intelligent:* A census taker asks a woman how many people live in her house and what their ages are. She tells him that her three daughters live in her house, that the product of their ages is 36, and that the sum of their ages is the number of the house next door. The census taker finds out the number of the house next door (13), but returns to tell the woman he does not have enough information. The woman tells him that her oldest daughter is sleeping upstairs. With this the census taker is able to figure out the daughters' ages. What are they?

Solution: 9, 2, 2. Thirty-six can only be broken into two sets of ages that equal 13 (9, 2, 2; and 6, 6, 1). When the mother says that the oldest one is sleeping, the solution is obvious.

Clark Kinnard includes the following traditional problem, which Sam Loyd is famous for: The combined ages of Mary and Ann are forty-four years. Mary is twice as old as Ann will be when Ann is three times as old as Mary was when Mary was three times as old as Ann.

Solution: Mary was $3x$ when Ann was $x$ (thus, $2x$ = difference); Mary was $5x$ when Ann was $3x$, and their ages totaled 44. $44/8x = 5\frac{1}{2}$. Mary = $27\frac{1}{2}$ yrs.; Ann = $16\frac{1}{2}$ yrs.

Charles W. Trigg includes the following in *Mathematical Quickies:* A man living in 1937 said he was $x$ years old in the year $x^2$. Also, the number of his years added to the number of his birth month equalled the square of the day of the month he was born. When was he born? Solution: May 7, 1892.

The following *weight problem* comes from Sam Loyd: Two turkeys weigh a total of 20 pounds. The smaller turkey sells for two cents more per pound than the larger turkey. The smaller turkey costs $.82, and the larger turkey costs $2.96. How much does each turkey weigh? Solution: 16 lbs. and 4 lbs.

The following weight problem comes from Martin Gardner, *Mathematical Puzzles:* If a basketball weighs $10\frac{1}{2}$ ounces plus half of its own weight, how much does it weigh? Solution: 21 ounces.

Sam Loyd includes the following *speed and distance puzzle:* A member of an exploring party to the North Pole decided to capture a bride for himself, following the custom of stealing a bride while she sleeps in her bearskin sack. He made the trip there at 5 mph and the trip back at 3 mph, taking exactly 7 hours for the entire trip. When he got back he found he had made a mistake and taken the bride's grandfather. How far did he travel?

Solution: $x$ = total number of miles traveled; $y$ = time it took to arrive; $z$ = time it took to return; $x/y = 5$; $x/z = 3$; $y + z = 7$; total distance = $29\frac{1}{4}$ miles.

Fixx includes a number of *sequence or series puzzles* in *More Games for the Super-Intelligent.* Here are three of them:

In what way are the following numbers arranged: 0, 2, 3, 6, 7, 1, 9, 4, 5, 8. Solution: by reverse alphabetical ordering of their first letters.

If water lilies double in area every twenty-four hours and there is one water lily on the lake at the beginning of summer and it takes sixty days for the lake to become completely covered with water lilies, on what day is it half covered? Solution: the fifty-ninth day.

List the next four numbers in the series: 12, 1, 1, 1, 2, 1, 3, . . . . Solution: 1, 4, 1, 5. The numbers represent the chimes on a clock which strikes once every half hour.

Gardner includes the following in *Mathematical Magic Show:* What is the missing number in the following series: 10, 11, 12, 13, 14, 15, 16, 17, 20, 22, 24, 31, 100, 10,000? Solution: 121; each number equals 16 in a different base numbering system, starting with base 16 and descending to base 2.

Fixx includes the following *grouping puzzle:* Divide the numbers 1–14 into the following groups:

I = 0, 3, 6, 8, 9

II = 1, 4, 7, 11, 14

III = 2, 5, 10, 12, 13

Which groups do the following numbers belong in: 15, 16, 17? Solution: 15 belongs in III (it consists of numbers with curved and straight lines); 16 belongs in III (numbers with curved and straight lines); 17 belongs in II (it consists of numbers with only straight lines).

The final example is a famous *train story problem:* A man walks to the railroad crossing at random times and keeps a record of which direction the first train that passes is headed. He knows that an equal number of trains is traveling in each direction; yet he finds that his records indicate a 3 to 1 ratio. Why?

Solution: The trains passing in one direction are quickly followed by the trains passing in the other direction, i.e., the trains going west come on the hour, the trains going east at 15 minutes after the hour. Thus, the first train is 3 times as likely to be from the west.

For additional recreations involving arithmetic and algebra refer to Calendars; Cryptarithms; Cryptography; Concealment Ciphers; Digital Problems; Guessing Numbers; Magic Squares; Number Patterns, Tricks, and Curiosities; Numerology; Substitution Ciphers; Transposition Ciphers, and Code Machines.

BIBLIOGRAPHY

Bakst, Aaron. *Mathematical Puzzles and Pastimes*. 2d ed. Princeton, N.J.: D. Van Nostrand Co., 1965.

―――. *Mathematics: Its Magic and Mastery*. New York: D. Van Nostrand Co., 1941.

Ball, W. W. Rouse. *Mathematical Recreations and Essays*. 1892. Rev., ed. H. S. M. Coxeter, London: Macmillan and Co., 1939.

Beiler, Albert H. *Recreations in the Theory of Numbers: The Queen of Mathematics Entertains*. 2d ed. New York: Dover, 1966.

Dudeney, Henry Ernest. *Amusements in Mathematics*. 1917. Reprint. New York: Dover, 1958.

―――. *The Canterbury Puzzles*. 1919. Reprint. New York: Dover, 1958.

Fixx, James F. *More Games for the Super-Intelligent*. New York: Doubleday, 1976.

―――. *Solve It!: A Perplexing Profusion of Puzzles*. New York: Doubleday, 1978.

Friend, J. Newton. *More Numbers: Fun and Facts*. New York: Charles Scribner's Sons, 1961.

―――. *Still More Numbers: Fun and Facts*. New York: Charles Scribner's Sons, 1964.

Gardner, Martin. *Mathematical Magic Show*. New York: Alfred A. Knopf, 1977.

―――. *Mathematical Puzzles*. New York: Thomas Y. Crowell Co., 1961.

Heafford, Philip. *The Math Entertainer: Teasers, Ticklers, Twisters, Traps and Tricks*. New York: Vintage Books, 1983.

James, Glenn, ed. *The Tree of Mathematics*. California: The Digest Press, 1957.

Kordemsky, Boris A. *The Moscow Puzzles*. Trans. by Albert Parry; ed. by Martin Gardner. New York: Charles Scribner's Sons, 1972.

Loyd, Sam. *Sam Loyd's Cyclopedia of Puzzles*. Reprint. *Mathematical Puzzles of Sam Loyd, Vols. I and II*. Ed. by Martin Gardner. New York: Dover, 1959, 1960.

O'Beirne, T. H. *Puzzles and Paradoxes*. New York: Oxford University Press, 1965.

Phillips, Herbert. *My Best Puzzles in Logic and Reasoning*. New York: Dover, 1961.

Simon, William, *Mathematical Magic*. New York: Charles Scribner's Sons, 1964.

Trigg, Charles W. *Mathematical Quickies*. New York: McGraw-Hill, 1967.

**ARITHMETIC RESTORATION.** See CRYPTARITHMS, CRYPTOGRAPHY, CONCEALMENT CIPHERS, SUBSTITUTION CIPHERS, TRANSPOSITION CIPHERS, AND CODE MACHINES.

**(THE) ARROW.** See ZENO'S PARADOXES.

**AUSTIN'S DOG.** See ZENO'S PARADOXES.

**AUTOMORPHIC NUMBERS.** See NUMBER PATTERNS, TRICKS, AND CURIOSITIES.

# B

**THE BARBER'S PARADOX.** See LOGICAL PARADOXES.

**BEASTING THE MAN.** See NUMEROLOGY.

**BERGEN.** See DOMINOES.

**BILLIARD CHESS.** See FAIRY CHESS.

**BLACK'S MACHINE.** See ZENO'S PARADOXES.

**BLOCK.** See DOMINOES.

**BORDERED MAGIC SQUARES.** See MAGIC SQUARES.

**BOSS.** See THE FIFTEEN PUZZLE.

**BRUSSELS SPROUTS.** See TOPOLOGY.

**BULO.** See NIM.

# C

**CAESAR CIPHER.** See CRYPTARITHMS, CRYPTOGRAPHY, CONCEAL-MENT CIPHERS, SUBSTITUTION CIPHERS, TRANSPOSITION CIPHERS, AND CODE MACHINES.

**CALENDARS** are the source of numerous mathematical pastimes.

Here are two calendar problems included in W. W. Rouse Ball's *Mathematical Recreations and Essays*. He attributes the first to E. Fourrey. It goes as follows:

Between 1725 and 1875, the French fought and won a battle on April 22 of some year, and 4,382 days later, also on April 22, fought and won another battle. The sum of the digits of the two years is 40. What were the dates of the two battles (the Gregorian calendar is employed)?

Since 4382 = 12 x 365 + 2, the date of the second battle must have been 12 years after that of the first battle. Only two leap years intervened; therefore, the year 1800 must be within the dates, meaning they must have occurred between 1788 and 1812. Of these years, only 1796 and 1808 produce 40 as a sum of their digits. Thus, the battles were fought on April 22, 1796 (Mondovi under Napoleon), and April 22, 1808 (Ecknugl under Davoust).

The second problem requires showing that the first or last day of every alternate century must be a Monday. This can be done by choosing any assigned date and combining that with the fact that the Gregorian cycle is completed in 400 years (20,871 weeks). The same process can be employed to prove that Friday is more likely to fall on the thirteenth day of the month than is any other day of the week.

The above calendar problems are easy to solve if the workings of the Gregorian calendar are understood. For instance, in order to solve the first problem, it is necessary to know that the Gregorian calendar establishes that the final year of each century is not a leap year, unless the number of the century is divisible by 4 (e.g., 1700, 1800, 1900, and 2100 are not leap years, but 2000 is a leap year).

*Calendar* is derived from the Latin *calendarium* (which means interest register or account book), itself a derivation from *calendae* or *kalendae* (which means the first day of the Roman month). The first practical calendar was developed by the Egyptians, and this calendar, in turn, was developed by the Romans into the Julian calendar, which was not replaced by the Gregorian calendar until the late sixteenth century.

The ancient Egyptian calendar was based on the moon (i.e., it was a lunar calendar) and was regulated by means of the star Sirius (which resulted in a year 12 minutes shorter than a solar year). Since the moon's motion around the earth is not an equal divisor of the 365.242199 day solar year, it was necessary for the Egyptians to devise a civil year of 365 days divided into three seasons, each 4 months of 30 days, with a 5 day intercalary period added to equal 365 days.

The lunar calendar was used to regulate religious affairs, and the civil calendar was used for government and administration. However, since the lunar and civil calendars kept getting out of sync, new lunar calendars were introduced (e.g., a 25 year cycle of intercalation). Since the old lunar calendar was still retained for agriculture, three calendars were being employed at the same time.

The early Roman calendar, supposedly drawn up by Romulus (c. seventh or eighth century B.C.), began the year in March and consisted of 10 months, 6 of 30 days and 4 of 31 days, a total of 304 days, and ended in December (leaving about a 2 month gap). Numa Pompilius (c. 715–673 B.C.), the second king of Rome, is said to have added January and February, resulting in a 12 month, 355 day year. The Etruscan Tarquinius Priscus (616–579 B.C.), the fifth king of Rome, is given credit for introducing the Roman Republican calendar, which still had 355 days: February had 28; March, May, Quintilis (July), and October had 31; January, April, June, Sextilis (August), September, November, and December had 29. To prevent it from getting too far out of step with the seasons, an intercalary month, Intercalans or Mercedonius, of 27 or 28 days was added every other year between February 23 and 24, generally omitting the final 5 days of February.

The Julian calendar, in an attempt to reconcile the lunar and solar schemes, was developed out of the Roman Republican calendar and the Egyptian calendar. It was initiated by Julius Caesar in the mid first century B.C., and an Alexandrian astronomer, Sosigenes, was given the task of devising it. Sosigenes abandoned the lunar calendar and arranged the months on a seasonal basis (as in the Egyptian calendar), with the length of 365.25 days per year (more accurate than the Egyptian figure of 365). In order to get the calendar date back in line with the equinoxes, two additional months (in addition to the standard intercalation) were inserted for the year 46 B.C. In order to achieve a 365.25 day year, Caesar instituted a 365 day year, with an extra day inserted between February 23 and 24 every fourth year. This was often misinterpreted (the fourth year of one period being considered an overlapping year with the first year of the next four year period) until Augustus corrected the error by omitting intercalary days between 8 B.C. and A.D. 8, when the true Julian calendar finally was properly established.

The official introduction of the seven day week did not take place until Emperor Constantine I in the fourth century.

The Julian calendar year, however, once again, was not accurate enough. The 365.25 day year seems accurate, but the real solar year is 365.242199 days. This is a difference of 11 minutes, 14 seconds a year; about $1\frac{1}{2}$ days in two centuries; and 7 days in 1,000 years. By 1545, the vernal equinox (used for determining Easter) was 10 days off its proper date. When the Council of Trent met the following December, it authorized Pope Paul III to correct the error. It was not until Gregory XIII became pope (1572), however, that any action was taken. Gregory assigned the task to Jesuit astronomer Christopher Clavius, who used the suggestions of Luigi Lilio.

It resulted in a papal bull in 1582, which simply eliminated 10 days and established a year of 365.2422 days. This year differed from the Julian by approximately 3.12 days every four centuries. It was thus established that three out of every four centennial years would not be leap years (i.e., that only centennials divisible by 4 would be leap years).

The Gregorian calendar continues to be generally used today. However, there have been serious proposals for more effective calendars. The international fixed calendar is a perpetual Gregorian calendar in which the year is divided into 13 months, each of 28 days, with an additional day added at the end. A new month, "Sol," is added between June and July. Neither the leap year day nor the additional day at the end of the year is attached to any week or month. Under this calendar every month begins on a Sunday and ends on a Saturday.

The world calendar attempts to divide the year up into equal quarters of 91 days, with an additional day at the end of the year. The first month of each quarter is 31 days, and the other two are 30 days. Once again, the extra day and leap year days are not included in weekly or monthly designations.

A perpetual calendar is one that allows the finding of the day of the week that any date fell on or will fall on, or in reverse, the months that begin on Sunday of any year, and so on. Karl Friedrich Gauss (1777–1855) gave the following formula for finding the day of the week corresponding to a given date:

$$W \equiv D + M + C + Y \text{ (modula 7)}$$

W = the day of the week, starting with Sunday

D = the day of the month

$M$ = the number assigned the month (see following table)

$C$ = the number assigned the century (see following table)

$Y$ = the number assigned the year (see following table)

If the month is January or February, the value of $Y$ must be diminished by 1. The following three tables indicate the values for $M$, $C$, and $Y$, respectively:

| Month | M |
|---|---|
| January | 0 |
| February | 3 |
| March | 3 |
| April | 6 |
| May | 1 |
| June | 4 |
| July | 6 |
| August | 2 |
| September | 5 |
| October | 0 |
| November | 3 |
| December | 5 |

| First Two Digits of the Year | C |
|---|---|
| Gregorian Calendar | |
| 15, 19, 23 | 1 |
| 16, 20, 24 | 0 |
| 17, 21, 25 | 5 |
| 18, 22, 26 | 3 |
| Julian Calendar | |
| 00, 07, 14 | 5 |
| 01, 08, 15 | 4 |
| 02, 09, 16 | 3 |
| 03, 10, 17 | 2 |
| 04, 11, 18 | 1 |
| 05, 12, 19 | 0 |
| 06, 13, 20 | 6 |

| **Last Two Digits of the Year** | **Y** | | |
| --- | --- | --- | --- |
| 00, 06, 17, 23, 28, 34, 45 | 0 | 51, 56, 62, 73, 79, 84, 90 | 0 |
| 01, 07, 12, 18, 29, 35, 40, 46 | 1 | 57, 63, 68, 74, 85, 91, 96 | 1 |
| 02, 13, 19, 24, 30, 41, 47 | 2 | 52, 58, 69, 75, 80, 86, 97 | 2 |
| 03, 08, 14, 25, 31, 36, 42 | 3 | 53, 59, 64, 70, 81, 87, 92, 98 | 3 |
| 09, 15, 20, 26, 37, 43, 48 | 4 | 54, 65, 71, 76, 82, 93, 99 | 4 |
| 04, 10, 21, 27, 32, 38, 49 | 5 | 55, 60, 66, 77, 83, 88, 94 | 5 |
| 05, 11, 16, 22, 33, 39, 44, 50 | 6 | 61, 67, 72, 78, 89, 95 | 6 |

Plugging in a date, say, June 21, 1986, Gauss' formula works as follows:

$$W \equiv 21 + 4 + 1 + 2 \text{ (modula 7)}$$
$$W \equiv 28 \times \tfrac{1}{7} \equiv 4.0 \equiv \text{Saturday}$$

If the remainder equals 0, then, in modula 7, it equals 7. Here is one more example (remember, it is the remainder that determines the day): July 4, 1986:

$$W \equiv 4 + 6 + 1 + 2 \text{ (modula 7)}$$
$$W \equiv 13 \times \tfrac{1}{7} \equiv 1.6 \equiv \text{Friday}$$

To discover what months of any year begin on, say, a Sunday, the same formula is applied to the first day of each month in turn, and whenever the first digit in the remainder comes up a 1, the month begins on a Sunday. January 1986 would be determined as follows:

$$W \equiv 1 + 0 + 1 + 1 \text{ (remember the rule for January and February) (modula 7)}$$
$$W \equiv 3 \times \tfrac{1}{7} \equiv .4 \ldots \equiv \text{Wednesday}$$

January does not begin on a Sunday; it begins on a Wednesday.

For similar mathematical recreations refer to Number Patterns, Tricks, and Curiosities.

BIBLIOGRAPHY

Bakst, Aaron. *Mathematical Puzzles and Pastimes*. 2d ed. Princeton, N.J.: D. Van Nostrand Co., 1965.
Ball, W. W. Rouse. *Mathematical Recreations and Essays*. 1892. Rev. by H. S. M. Coxeter. London: Macmillan and Co., 1939.
———. *A Short Account of the History of Mathematics*. 1888. Reprint. London: Macmillan and Co., 1935.
Collins, A. Frederick. *Fun with Figures*. New York: D. Appleton and Co., 1928.

Kline, Morris. *Mathematics in Western Culture*. New York: Oxford University Press, 1953.

Kraitchik, Maurice. *Mathematical Recreations*. 2d rev. ed. New York: Dover, 1953.

Ronan, C. A. "Calendar." In *Macropaedia: Encyclopedia Britannica,* 1985.

Simon, William. *Mathematical Magic*. New York, 1964.

**CANTORISM.** See PARADOXES OF THE INFINITE.

**CARD FRAME PUZZLE.** See CARD TRICKS AND PUZZLES.

**CARD TRICKS AND PUZZLES** are mathematical recreations involving the use of standard playing cards.

Henry Ernest Dudeney includes the following card puzzles in *Amusements in Mathematics:* The first is titled "The Card Frame Puzzle." The ace to ten of diamonds are arranged as follows:

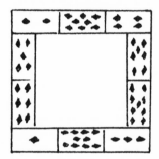

The object is to rearrange each row and column so that they add up to the same number:

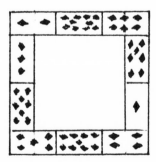

The second is titled "The Cross of Cards." The ace through nine of diamonds are arranged as follows:

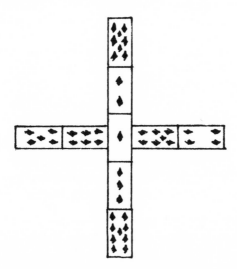

The object is to rearrange the cards (substituting at least one card from the vertical bar with at least one card from the horizontal bar each time) to create as many different arrangements as possible where both the vertical total and the horizontal total are equal. The arrangements on the horizontal bar may consist of the following (the rest of the cards naturally are used on the vertical bar): 56174, 35168, 34178, 25178, 25368, 15378, 24378, 14578, 23578, 24568, 34567, 14768, 23768, 24758, 34956, 24957, 14967, 23967.

The third is titled "The 'T' Card Puzzle." The cards are arranged as follows:

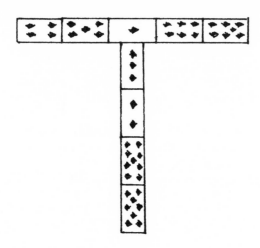

As with the previous puzzle, the object is to come up with the greatest number of possible arrangements where the vertical and horizontal rows both add up to the same number (excluding mirror inversions). There are 10,368 possibilities.

Here are two of the card tricks included in Maurice Kraitchik's *Mathematical Recreations:* In the first, $n$ red cards and $n$ black cards are placed alternately in a row, leaving two empty spaces at one end. The object is to separate the red cards from the black cards by moving two adjacent cards at a time into the vacated spaces in $n$ number of moves. The following is the solution for $n = 4$:

$$. . R_1B_1R_2B_2R_3B_3R_4B_4$$
$$B_3R_4R_1B_1R_2B_2R_3 . . B_4$$
$$B_3R_4R_1B_1 . . R_3R_2B_2B_4$$
$$B_3 . . B_1R_4R_1R_3R_2B_2B_4$$
$$B_3B_2B_4B_1R_4R_1R_3R_2 . .$$

In the second, the object is to guess a selected card. Someone chooses (in his mind) one of 27 (or 21) cards in a pack. The person guessing then shuffles the pack and deals it into three piles (one card at a time). The chooser is then asked which pile the card is in. The dealer then places that pile between the other two piles and (without shuffling) deals three piles as before. Once again, the chosen pile is placed between the other two, and the process is repeated. This time, when the pile is chosen, the dealer can assert that the chosen card is the center one in that pile.

For similar recreations refer to Guessing Numbers; Magic Squares.

BIBLIOGRAPHY

Dudeney, Henry Ernest. *Amusements in Mathematics.* 1917. Reprint. New York: Dover, 1958.

Kraitchik, Maurice. *Mathematical Recreations.* 2d rev. ed. New York: Dover, 1953.

**CARROLL'S CLOCKS.** See LOGICAL PARADOXES.

**CASTING OUT NINES.** See GUESSING NUMBERS.

**CENTURION** is an activity that combines language and mathematics.

All of the letters of the alphabet are given successive numerical values (i.e., A = 1, B = 2, C = 3, and so on).

The play begins by having one player write down a word whose combined numerical value is 10 or less, e.g.:

$$ACE = 1 + 3 + 5 = 9$$

The second player then writes another three-letter word, this word beginning with the final letter of the first word, and indicates the total of all of the letters so far, e.g.:

EAT = 26 (5 + 1 + 20) + 9 = 35

Play continues in this manner until a three-digit total (100 or more) is reached. The player who reaches it "bursts" and loses.

Double Centurion can be played by three or more players with the burst score upped to at least 200 points.

For similar mathematical word play refer to ABC Words; ACE Words; Numwords. For mathematical mysticism involving letters and words refer to Numerology.

BIBLIOGRAPHY

Parlett, David. *Botticelli and Beyond: Over 100 of the World's Best Word Games.* New York: Pantheon Books, 1981.

**CHECKERS,** also called draughts, is a game played on the thirty-two black or white squares of a standard chess board (*see* Chess) by two opponents, each having twelve identical men. The object of the game is either to capture all of the opponent's men or block them so that they cannot move.

The pieces are arranged on the board as follows:

The squares on a checker board are either designated in the same manner as those on a chess board or as follows:

| | 32 | | 31 | | 36 | | 29 |
|---|---|---|---|---|---|---|---|
| 28 | | 27 | | 26 | | 25 | |
| | 24 | | 23 | | 22 | | 21 |
| 20 | | 19 | | 18 | | 17 | |
| | 16 | | 15 | | 14 | | 13 |
| 12 | | 11 | | 10 | | 9 | |
| | 8 | | 7 | | 6 | | 5 |
| 4 | | 3 | | 2 | | 1 | |

The black pieces are placed on squares 1–12, the white pieces on squares 21–32. Black moves first. The pieces are moved forward diagonally one space at a time. If an opponent's piece is blocking a move and the square beyond that piece is vacant, the piece *must* be captured by jumping over it to the vacant square beyond. This jumping *must* be continued as long as possible. The captured men are removed from the board.

If a player does not make a capture when it is possible, the opponent has the right to remove the player's piece from the board before making his move (this is called *huffing*) or to simply force the opponent to retract his move and make the necessary capture.

When a checker reaches the opposite side of the board, it is *crowned* a king (indicated by placing a second checker on top of it). This king may then move backward as well as forward. On the move it makes to become a king, however, it may not move any farther.

When one player has captured all of the opponent's players or made it impossible for the opponent to move, he wins. If neither side is able to do so, the game is a draw.

In France the standard game, called *jeu des dames* (game of the queens), varies from the standard American game as follows: A piece is called a pawn until it reaches the opponent's edge of the board, when it becomes a queen. The queen may then move along its diagonals any distance it wishes, over any pieces in its way, capturing those pieces over which it passes. In Poland, a 10 × 10 board is used, with twenty men to a side. The other rules are the same as in the French game.

*Cheskers* a hybrid game combining both checkers and chess, was invented by Solomon W. Golomb in 1947. It is played on the thirty-two black squares of a standard checker board. One piece, the cook, is a modified knight that moves three spaces forward and one to either side. It also has the ability to move over intervening pieces as a knight does in chess. Each player also has two kings, which move as do kings in chess; a bishop, which moves as does a bishop in chess; and eight men, which move as checkers move in checkers.

The men and the kings capture as do checkers in a standard game of checkers. The bishops and cooks capture as in chess (by moving onto the victim's square). If a checker capture is possible, it must be made, unless a chess capture is also possible, in which case the chess capture may be made instead. If only a chess capture is possible, it is optional.

The original set-up of the pieces is as follows:

When a player gets a man to the opponent's side of the board he chooses whether that man becomes a king, bishop, or cook. Players alternate moves. The object is to capture both of (all of) the opponent's kings. If one player cannot move, he loses.

Such variations on standard checkers suggest the close relationship checkers has with chess. It is also possible to construct checker problems similar to chess problems, though checker problems are limited. For similar recreations refer to Chess; Chess Problems; Fairy Chess.

BIBLIOGRAPHY

Gardner, Martin. *Mathematical Magic Show*. New York: Alfred A. Knopf, 1977.

Kraitchik, Maurice. *Mathematical Recreations*. 2d rev. ed. New York: Dover, 1953.

Lasker, Edward. *Chess and Checkers: The Way to Mastership*. 3d rev. ed. New York: Dover, 1960.

**CHESKERS.** See CHECKERS.

**CHESS,** also called the Royal Game, has been the most popular positional game in the entire field of mathematical games, puzzles, and amusements. Not only has the standard game of chess provided for numerous mathematical explorations (mostly geometrical), but the chessboard and the chessmen have been a source of continuous mathematical play (*see* Chess Problems; Fairy Chess).

The English name of the game, *chess,* along with the French name, *échecs,* and the Italian name, *scacchi,* can be traced through the Latin plural of *scaci* (*scachi, scacci,* meaning "chessmen") to the Arabic and Persian name of the chess king, *shāh.* The term "checkmate" also traces back to the Persian *shāh* ("king") *māt* ("dead"). Persian literature, in turn, generally attributes the game to India and connects its introduction to Persia with the book *Kalila wa Dimna,* A.D. 531–578.

While the facts of the origin and spread of chess cannot be precisely determined, there is general agreement that it did originate in either India or China, certainly in a far different form than we know it today, sometime before the sixth century. There are numerous Muslim legends of its origin, interesting mainly as curiosities. In one of them (found in H. J. Murray, *A History of Chess*) Hashrān, an Indian monarch, asked Qaflān, a sage, to create a game that would symbolize man's dependence on destiny and fate. The sage invented the game of Nard. The board stood for a year. It had twenty-four houses, because there are twenty-four hours in a day. It was arranged in two halves of twelve houses each to symbolize the twelve months of the year or the twelve signs of the zodiac. The number of pieces was thirty to symbolize the thirty days of a month. The two dice stood for day and night. Their faces were arranged so that the six stood opposite the one, the five opposite the two, the four opposite the three, so that the total of each was seven, the number of days in a week and the seven components of the solar system (i.e., the Sun and Moon, and the planets Saturn, Jupiter, Mars, Mercury, and Venus). Each player's move was determined by a throw of the dice, thus symbolizing man's dependence on fate. Hashrān was thrilled, and the first form of chess was invented.

Balhait, a later king, was informed by a Brahman that this game was anti-religious. Balhait and his Brahman friend then proceeded to change the game so that it would demonstrate the qualities of prudence, diligence, thrift, and knowledge, and therefore oppose man's dependence on fate. In this game, the board was eight squares by eight squares, and there were sixteen pieces on each side (a *shāh,* a *firz,* two *fīls,* two *faras,* two *rukhs,* and eight *pawns*). It was modeled after war on the belief that war is the most effective place to learn the value of administration, decision, prudence, caution, arrangement, strategy, circumspection, vigor, courage, force, endurance, and bravery.

Whatever its true origin, chess appears to have traveled from India through

Persia, and reached Spain and southern Italy by way of the Muslim empire by the eleventh century. During the twelfth and thirteenth centuries it began to evolve into the modern form. The moves of the *baidaq* (pawn), *shah* (king) and *fers* (queen) were extended. At the end of the fifteenth century the *fers* moves were extended to modern form (at first this piece was called the "mad" or "furious" queen). At the same time the *alfil* (bishop), which previously had been limited to a single diagonal, had its moves extended to those of the modern bishop. In the sixteenth century, "castling" was introduced. Since then the standard European form of chess has remained constant (variations exist under the title of Fairy Chess and in various corners of the world, e.g., Japan and China).

In the standard game of chess, two players face each other across a black and white checkered board of sixty-four squares, so placed that a white square is on the extreme right of each player. Each player has sixteen pieces (or men): a king, a queen, two bishops, two knights, two rooks (or castles), and eight pawns. The rooks are placed on the corner squares, the knights on the next square inside of the rooks on the end row, the bishops inside the knights, the queen inside the bishop and on the square of the queen's color, and the king inside the bishop and on the square of the opponent's color. The eight pawns are lined up on the second row in, in front of the other pieces:

The object of the game is to place the opponent's king in checkmate. A king is in checkmate when a player cannot move his king to a position where his opponent will be unable to capture him in the next move. In other words, when a player puts his opponent in check (and he must announce it), the opponent must move out of check in his move or lose the game.

The chess pieces move as follows:

The *king* may move one space in any direction, provided it does not move into check or onto a space occupied by one of its own men.

The *queen* may move as far in any direction as wished, provided it does not change direction or find its progress blocked.

The *bishop* may move diagonally in the same manner as the queen.

The *knight* is the only piece that may jump over other pieces. It moves either one space foward or backward and two to the left or right, or two spaces forward or backward and one space to the left or right (an L-shaped move).

The *rook* moves either forward or backward or to either side in the same manner as a queen.

The *pawn* moves only forward, one space at a time, except on its first move, when it may move two spaces forward. If another piece is in front of it, a pawn is blocked. It captures any piece to the diagonal in a forward movement of one space. A pawn may be promoted to any rank except king if it reaches the opposite side of the board.

If a player has no move except to move his king into check, a stalemate is reached (a tie). The game may also end in a tie for the following reasons: the inability of either player to achieve a checkmate; perpetual check; the triple recurrence of an identical position with the same player having the move; the absence of material exchange or a pawn move within a sequence of fifty moves; and/or a mutual agreement by both players.

Castling is a compound move of both the king and one of the rooks. The king moves two squares toward the rook, and the rook jumps over the king and rests on the other side of it. It may be performed only once by a player and only if neither the king nor the rook has previously been moved. Also, the intervening squares must be vacant, the two squares on the side of the king where the castling is to take place must not be under the command of the opponent's pieces, and the king must not be in check.

If a pawn makes a double-space move for its first move, an opponent's pawn that could have captured it had it made only a one-space move may capture it *en passant* (i.e., in passing) on the next move only.

In order to discuss chess and its various moves and problems, a system of notation has been developed. Across the files or columns from left to right the squares are designated a, b, c, d, e, f, g, and h. The forward and backward ranks, rows, or columns are designated 1, 2, 3, 4, 5, 6, 7, and 8. The pieces are then designated as K (king), Q (queen), B (bishop), Kt (knight), R (rook), and P (pawn), thus, QB is queen's bishop, etc.

The white king would begin the game at Ke1. If the king's pawn were to move two spaces forward, it would be indicated as eP4. If there is ambiguity, the space from which the piece moves is indicated in parentheses, e.g., eP(e2)e4. If the piece makes a capture, the square moved to is omitted and replaced by "x" (takes). Thus, Kt x R means knight takes rook. If the move results in a check, + or *ck* is placed after the move. 0-0 means a castle on the king's side; 0-0-0 means a castle on the queen's side.

The World Chess Federation, officially the Federation Internationale des Echecs (FIDE), is the generally accepted international governing body. It is in charge of interpretation and revision of rules, and it holds or supervises the International Team Tournament every two years and organizes the World Championship. The International Team Tournaments (also called the Chess Olympics, Olympiads, and World Team Championships) began informally in 1924. The first official tournament was held in London in 1927 with Hungary finishing first, Denmark second, and Great Britain third. When Alexander A. Alekhine died in 1946 officially undefeated, the World Chess Federation established an orderly determination of the title. Mikhail Botvinnik of the U.S.S.R. was the first champion in 1948. Since 1965 an organized three-year cycle of qualifying tournaments has been established to find a challenger for the reigning champion. A similar tournament has been established for women. The first winner, L. Rudenko (1950), was also from the U.S.S.R.

For an in-depth study of the history of chess refer to H. J. R. Murray, *A History of Chess*. For additional mathematical play based on chess *see* Chess Problems. For variations on standard chess *see* Fairy Chess. For an additional legend of the beginning of chess *see* The Tower of Hanoi.

BIBLIOGRAPHY

Ball, W. W. Rouse. *Mathematical Recreations and Essays*. 1892. Rev. by H.S.M. Coxeter. London: Macmillan and Co., 1939.

Dickins, Anthony. *A Guide to Fairy Chess*. New York: Dover, 1971.

Gardner, Martin. *Martin Gardner's Sixth Book of Mathematical Games from "Scientific American."* San Francisco: W. H. Freeman and Co., 1971.

———. *Wheels, Life and Other Mathematical Amusements*. San Francisco: W. H. Freeman and Co., 1983.

Kraitchik, Maurice. *Mathematical Recreations*. 2d rev. ed. New York: Dover, 1953.

Madachy, Joseph S. *Mathematics on Vacation*. New York: Charles Scribner's Sons, 1966.

Murray, H. J. R. *A History of Chess*. 1913. Reprint London: Oxford University Press, 1962.

Northrop, Eugene P. *Riddles in Mathematics: A Book of Paradoxes.* New York: D. Van Nostrand Co., 1944.

Schaaf, W. L. "Number Games and Other Mathematical Recreations." In *Macropaedia: Encyclopedia Britannica,* 1985.

Steinhaus, H. *Mathematical Snapshots.* New York: Oxford University Press, 1969.

**CHESS PIECE TOURS.** See CHESS PROBLEMS.

**CHESS PROBLEMS,** including fairy chess problems, are various mathematical amusements based on aspects of the chessboard or variations of it, chessmen or variations of them, and the game of chess or variations of it. They include endgames and tours of the board by the different pieces.

The endgame is an important part of the early literature on chess, both in Muslim and European literature. The Arabic terms for endgame positions are *manṣūba, manṣūbāt,* or *manaṣib,* the past form of the verb *naṣaba,* which means "to erect," "to appoint," or "to set up," in other words, an arrangement or problem. The major purpose of these problems seems to have been as exercises, as a form of practice, though some were undoubtedly admired for their cleverness.

Geoffrey Chaucer, in *Book of the Duchess,* has his hero, defeated at chess by Dame Fortune, say:

> But god wolde I had ones or twyes
> Y-koud and knowe the Ieupardyes
> That coude the Grek Pithagores!
> I shulde have pleyd the bet at ches,
> And kept my fers the bet therby. [Lines 665–669]

This passage shows that Chaucer, at least, regarded the problems mainly as a means of acquiring the skills necessary for the real game of chess. There is little other medieval literature on problems, though H. J. R. Murray has found an additional passage from J. Lydgate's *Troy Book.*

Murray (and others) was unable to find any manuscripts of chess problems earlier than 1250 but assumes that they were being puzzled over earlier than this date. At any rate, by the end of the fourteenth century the *Bonus Socius* and by the middle of the fifteenth century the *Civis Bononiae,* two major collections of chess problems, appear. It is at this time that the modern form of chess solidified, and variations on it can be classified under the designation of fairy chess (*see* Fairy Chess).

Between 1918 and 1958, such important endgame composers and collectors as

T. R. Dawson (1889–1951) promoted fairy chess and chess problems, bringing them to their highest levels of activity and sophistication. Here are some of the more interesting or "classical" problems:

The shortest possible game, sometimes called Fool's Mate, goes as follows (refer to Chess for interpretation of the symbols):

| White | Black |
|---|---|
| 1. P-KB3 (or 4) | 1. P-K3 |
| 2. P-KKt4 | 2. Q-KR5 (mate). |

Sam Loyd's shortest possible stalemate without a piece being lost:

| White | Black |
|---|---|
| 1. P-Q4 | 1. P-Q3 |
| 2. Q-Q2 | 2. P-K4 |
| 3. P-QR4 | 3. P-K5 |
| 4. Q-KB4 | 4. P-KB4 |
| 5. P-KR3 | 5. B-K2 |
| 6. Q-KR2 | 6. B-K2 |
| 7. R-QR3 | 7. P-QB4 |
| 8. R-KKt3 | 8. Q-QR4 (ch) |
| 9. Kt-Q2 | 9. B-KR5 |
| 10. P-KB3 | 10. B-QKt6 |
| 11. P-Q5 | 11. P-K6 |
| 12. P-QB4 | 12. P-KB5 |

Stalemate.

Chess tours involve placing one of the chess pieces on some square of the board and requiring that it move in some manner to solve a problem. The most common problem is to traverse the board, touching each square once and only once.

The first tour (discussed in Maurice Kraitchik, *Mathematical Recreations*) involves not a chessman but a checker man. The object is to form Pascal's triangle (*see* Number Patterns, Tricks, and Curiosities) in checker moves. Designate the cells of the chess board from 1 to 8 across the bottom and from 1 to 8 bottom to top (thus, the lower left-hand corner cell would be 1,1). The checker man moves one step at a time along a diagonal, always in such a manner that he changes rows in the same direction, thus allowing us to be concerned only with which column he occupies. If the piece starts at, say, (4,1), this cell can be marked with a (1) to indicate, perhaps arbitrarily, that he can only get to this square in one manner. The same, no longer arbitrarily, can be done for the two moves he can make from (4,1), i.e., (3,2) or (5,2). In the next move he can go from either (3,2) to (4,3) or (2,3), or from (5,2) to (6,3) or (4,3). This can be continued, producing the following Pascal's triangle:

```
  . . .    .    .    .    . . .
 1  7  21  35   35  21  7  1
  1  6   15  20  15   6  1
   1   5  10  10   5  1
    1   4   6   4  1
     1   3   3  1
      1   2  1
       1  1
        1
```

The most often discussed tours are the knight's tours. A knight's reentrant tour requires the knight to visit every square of the board once and end where it started. Following are four of the possible knight's reentrant tours (they are included in *Martin Gardner's Sixth Book of Mathematical Games from "Scientific American"*):

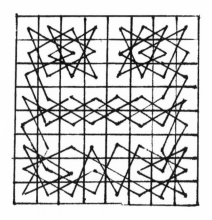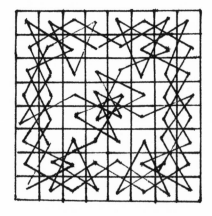

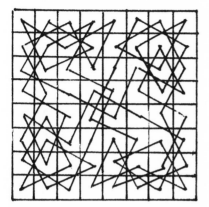 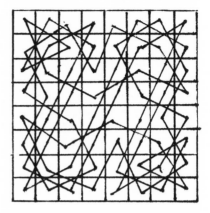

W. W. R. Ball, *Mathematical Recreations and Essays,* includes a solution given by D. Montmort and A. De Moivre at the beginning of the eighteenth century. The following is the same numerical solution:

| 34 | 49 | 22 | 11 | 36 | 39 | 24 | 1  |
|----|----|----|----|----|----|----|----|
| 21 | 10 | 35 | 50 | 23 | 12 | 37 | 40 |
| 48 | 33 | 62 | 57 | 38 | 25 | 2  | 13 |
| 9  | 20 | 51 | 54 | 63 | 60 | 41 | 26 |
| 32 | 47 | 58 | 61 | 56 | 53 | 14 | 3  |
| 19 | 8  | 55 | 52 | 59 | 64 | 27 | 42 |
| 46 | 31 | 6  | 17 | 44 | 29 | 4  | 15 |
| 7  | 18 | 45 | 30 | 5  | 16 | 43 | 28 |

Leonhard Euler, *Mémoires de Berlin* (1759), was the first seriously to attempt a mathematical analysis of the problem. He began by moving the knight over the board randomly, until it could no longer move (leaving a few cells uncovered). Say all of the cells except for four have been traversed. Make the path 1 through 60 a reentrant path (i.e., list the numbers of the possible cells that the first move can move into and the numbers of the possible cells 60 can move into). If any of the numbers that 1 can move into form a sequence with the numbers 60 can move into, then the sequence can be changed to move from 1 to 60 (i.e., if 1 can move into 52 and 60 can move into 51, then cells 1–51 and 60–52 form a reentrant

route of 60 moves). Thus, if the numbers 60–52 are replaced by 52–60, the steps
will be numbered consecutively.

Then add the four cells not yet touched to the route by taking them one at a
time, listing the cells they could be moved to or from, picking one of the cells,
and using that cell as the final cell of the route (i.e., say one of the untouched
cells could be reached from numbers 51, 53, 41, 25, 7, 5, and 3; choose one of
those cells, say, 51; make it the final cell of the route of 60 cells by adding 9 [the
difference between 51 and 60] to each of the cells in the diagram and replacing
61–69 by numbers 1–9). The same process can be used to make the tour a
reentrant tour.

Euler then proceeded to show how this same process could be used to solve
variations on the standard problem, e.g., the following half-board version:

| 58 | 43 | 60 | 37 | 52 | 41 | 62 | 35 |
| 49 | 46 | 57 | 42 | 61 | 36 | 53 | 40 |
| 44 | 59 | 48 | 51 | 38 | 55 | 34 | 63 |
| 47 | 50 | 45 | 56 | 33 | 64 | 39 | 59 |
| 22 | 7  | 32 | 1  | 24 | 13 | 18 | 15 |
| 31 | 2  | 23 | 6  | 19 | 16 | 27 | 12 |
| 8  | 21 | 4  | 29 | 10 | 25 | 14 | 17 |
| 3  | 30 | 9  | 20 | 5  | 28 | 11 | 26 |

The five queen's tour problems given below are included in *Martin Gardner's
Sixth Book of Mathematical Games from "Scientific American."* They all refer
to the following board:

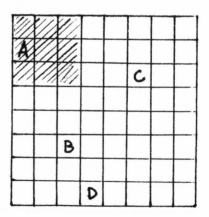

1. Start with the queen on square A and traverse all of the gray-shaded squares in four continuous moves.

2. Start with the queen on D, the starting place for the white queen in a standard chess set-up, and move the queen the farthest distance possible (measurable in inches, not squares) in five moves. The queen may not cross her own path or visit the same square twice. (This is one of the most common of the queen problems; Ball also includes it in his *Mathematical Recreations and Essays*.

3. Start the queen on cell B and pass through every cell only once in fifteen moves, ending on cell C.

4. Start with the queen in a corner cell. Traverse every cell of the board at least once and finish on the starting cell in just fourteen moves. (This is one of Sam Loyd's problems.)

5. Take a seven by seven cell board. Perform the same task as in problem 4 above in twelve moves.

Here are the solutions:

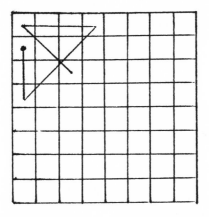

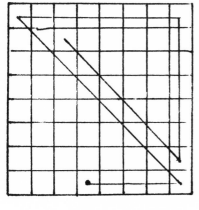
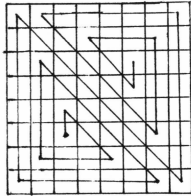

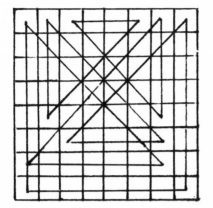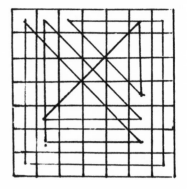

The first problem is simply another version of the geometric puzzle requiring the connecting of nine dots arranged as follows with four lines without lifting one's hand from the paper:

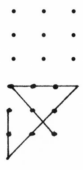

A more complicated version of this puzzle is the task of connecting sixteen dots with six lines without lifting one's hand from the paper:

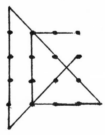

Two other problems involve placing queens on a chessboard. In one of these the object is to place eight queens on the board so that they command the fewest possible squares. If they are placed on cells 21, 22, 62, 71, 73, 77, 82, and 87, eleven cells are not under their command (this solution is given by Ball, and it is the best I have encountered).

The other problem is to place a number of queens less than five on a chessboard to command as many cells as possible. Once again, Ball has given the best solution I have encountered by placing the queens on cells 35, 41, 76, and 82.

H. Steinhaus, *Mathematical Snapshots,* includes the following example, which is neither a victory, a draw, nor "pat" (a situation which makes suicide necessary for one of the kings). It is White's move, but White cannot move:

|   |   |   |   | BK | WP | WR | WK |
|---|---|---|---|----|----|----|----|
|   |   |   |   | WP | BN | WP | WP |
|   |   |   |   |    |    |    |    |
|   |   |   |   |    |    |    |    |
|   |   |   |   |    |    |    |    |
|   |   |   |   |    |    |    |    |
|   |   |   |   |    |    |    |    |
|   |   |   |   |    |    |    |    |

Joseph S. Madachy, *Mathematics on Vacation,* includes a number of chess placement problems. Here are two examples:

1. What is the maximum number of queens that can be placed on a chessboard so that none of them can capture another (assuming they are all belligerent toward each other)? Answer: 8.
2. What is the maximum number of bishops that can be placed on a chessboard so that none of them can capture another (once again, assuming they are all belligerent toward each other)? Answer: 14.

There is a great deal of literature on chess problems; the scope of this book allows only a few examples to be cited. I end with the shortest possible game ending in perpetual check, once again first published by Sam Loyd (1866):

| White | Black |
|-------|-------|
| 1. P-KB4 | 1. P-K4 |
| 2. K-B2 | 2. Q-KB3 |
| 3. K-N3 | 3. QxP (check) |

For similar mathematical amusements refer to Chess; Fairy Chess. Refer to the various geometrical entries for similar geometrical play. Refer to H. J. R. Murray, *A History of Chess,* for detailed discussion of the history of chess problems.

BIBLIOGRAPHY

Ball, W. W. Rouse. *Mathematical Recreations and Essays.* 1892. Rev. by H. S. M. Coxeter. London: Macmillan and Co., 1939.

Dickens, Anthony. *A Guide to Fairy Chess.* New York: Dover, 1971.

Gardner, Martin. *Martin Gardner's Sixth Book of Mathematical Games from "Scientific American."* San Francisco: W. H. Freeman and Co., 1971.

————. *Wheels, Life and Other Mathematical Amusements.* San Francisco: W. H. Freeman and Co., 1983.

Kraitchik, Maurice. *Mathematical Recreations.* 2d rev. ed. New York: Dover, 1953.

Madachy, Joseph S. *Mathematics on Vacation.* New York: Charles Scribner's Sons, 1966.

Murray, H. J. R. *A History of Chess.* 1913. Reprint London: Oxford University Press, 1962.

Schaaf, W. L. "Number Games and Other Mathematical Recreations." In *Macropaedia: Encyclopedia Britannica,* 1985.

Steinhaus, H. *Mathematical Snapshots.* New York: Oxford University Press, 1969.

**CHINESE RINGS** are an arithmetical puzzle, a toy.

A fixed number of rings (generally from five to eight) are strung on a bar so that only the ring at the extreme right end may be taken off or put on the bar without additional manipulations. All other rings can only be taken off or put on when the ring immediately to the right of each is on and all others to the right are off. The rings are in fixed order. Except for the two rings at the extreme right, which can be taken off or put on together, all of the rings must be taken off or put on one at a time.

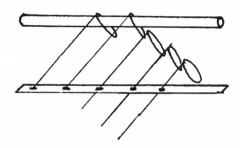

The rings, then, may be taken off one at a time or two at a time. On a toy with five rings, let "1" indicate that the ring is on the bar in order and let "0" indicate that the ring is off the bar in order. The representation for all of the rings on the bar would then be: 11111. Three moves are possible. The ring on the far right may be removed, leaving: 11110. The second ring from the far right may be removed, leaving: 11101. The two rings on the far right may be removed together, leaving: 11100. If the first ring were removed (11110) it would not be possible to move to 11101 in one move.

There is a simple rule for counting the number of steps needed to move from one position to another. (Maurice Kraitchik gives M. Gros credit for its discovery.) To determine the number of steps to move from one position to another on a toy with $n$ number of rings (say, five) an $n$-digit (e.g., five-digit) binary notation can be set up. First, starting at the left, for those rings *on* the bar, assign alternate 1's and 0's (thus, a bar of five rings with all five on it would be represented as 10101). If the ring to the far left is *off* the bar, it is assigned a 0. If any other ring is *off* the bar, it is assigned the same digit the ring to its left has. Thus, Gros notation for a set-up of rings of 11011 would be 1 (the left ring is *on* the bar), 0 (an alternating 0 for the second ring from the left, because it is also *on* the bar), 0 (a digit corresponding to the digit to its left, since it is *off* the bar), 1 (an alternating 1, because it is *on* the bar), and 0 (an alternating 0, because it is *on* the bar): 10010.

Applying Gros notation, the set of all possible positions corresponds to the set of all numbers from $0$ to $2^n-1$, inclusive. Since the object of the puzzle is to move from 10101 . . . to 00000 . . . in the fewest possible moves, the mathematical formula is $\frac{1}{3}(2^{n+1}-2)$, if $n$ is even, and $\frac{1}{3}(2^{n+1}-1)$, if $n$ is odd, if the rings are moved one at a time. If the rings are moved two at a time, it is possible to save two steps out of every eight, except at the finish, where one step out of every two is saved if $n$ is even, and one step out of five is saved if $n$ is odd, resulting in $2^{n-1}-1$ if $n$ is even, and $2^{n-1}$ if $n$ is odd.

In *Mathematics: Problem Solving Through Recreational Mathematics* Bonnie Averbach and Orin Chein include the following problem based on the same concepts as Chinese rings:

In order to prevent tampering by unauthorized individuals, a row of switches at a defense installation is wired so that, unless the following rules are followed in manipulating the switches, an alarm will be activated:

1. The switch on the right may be turned on or off at will.

2. Any other switch may be turned on or off only if the switch to its immediate right is on and all other switches to its right are off.

What is the smallest number of moves in which such a row of switches, which are all on, may be turned off without activating the alarm if:

(a) There are three switches in the row?

(b) There are four switches in the row?

(c) There are five switches in the row?

(d) There are six switches in the row?

This problem is equal to the rings problem, if the rings are moved one at a time. So the formula $\frac{1}{3}(2^{n+1}-2)$ if $n$ is even, and $\frac{1}{3}(2^{n+1}-1)$ if $n$ is odd, applies. Thus, the answers are: (a) 5; (b) 10; (c) 21; (d) 42.

According to Maurice Kraitchik, *Mathematical Recreations,* the theory of Chinese rings suggested the idea of the Tower of Hanoi to its creator, Edouard Lucas (*see* The Tower of Hanoi).

BIBLIOGRAPHY

Averbach, Bonnie, and Orin Chein. *Mathematics: Problem Solving Through Recreational Mathematics.* San Francisco: W. H. Freeman and Co., 1980.
Kraitchik, Maurice. *Mathematical Recreations.* 2d rev. ed. New York: Dover, 1953.

**CIPHER.** See CRYPTARITHMS, CRYPTOGRAPHY, CONCEALMENT CIPHERS, SUBSTITUTION CIPHERS, TRANSPOSITION CIPHERS, AND CODE MACHINES.

**CLOCK PUZZLES.** See ARITHMETIC AND ALGEBRAIC PROBLEMS AND PUZZLES.

**CODE.** See CRYPTARITHMS, CRYPTOGRAPHY, CONCEALMENT CIPHERS, SUBSTITUTION CIPHERS, TRANSPOSITION CIPHERS, AND CODE MACHINES.

**CODE MACHINES.** See CRYPTARITHMS, CRYPTOGRAPHY, CON-CEALMENT CIPHERS, SUBSTITUTION CIPHERS, AND TRANSPOSI-TION CIPHERS.

**COIN PROBLEMS AND PUZZLES** are those recreations based on the properties of coins.

Cube-O is a coin game (a form of Nim; *see* Nim) invented by Maxey Brooke, *Fun for the Money: Puzzles and Games with Coins.* Nine stacks of three coins each or sixteen stacks of four coins each are piled and brought together in the shape of a cube. Two players alternately remove the coins according to the following rules: 1) the players must take the top coin of any stack; 2) the players may take any two, three, or four coins in a row, stack, or file (not a diagonal) provided they touch each other; 3) the player who takes the final coin loses.

Peg solitaire is a coin puzzle where the coins are set up in a triangle (similar to the arrangement of bowling pins). The object is to reduce the number of coins to a single coin by jumping them. No sliding is allowed. The rules are the same as those for checkers (*see* Checkers). Jumps may be made in six possible directions. Continuous chains of jumps count as only one move. Five moves is the minimum for ten coins; nine for fifteen.

A coin problem involves setting up ten or fifteen coins in a triangle similar to peg solitaire and then inverting the triangle by sliding one penny at a time into a new position in which it touches two other pennies. It can be done in three moves for ten coins, five moves for fifteen.

Coins can also be used for tree-planting problems, such as the following: A farmer wishes to plant nine trees so that they form ten straight rows of trees with three trees in each row:

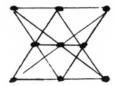

Coins are also used for rotation problems, such as the following: How many rotations does it take for a penny to roll around another penny? Solution: Two. The rotation of a penny around any closed chain of pennies (it must touch each penny) is $\frac{2}{3}n + 2$ rotations ($n =$ the number of pennies). The rotation of a penny around the inside of a closed chain equals $(n/3) - 2$.

Here is another penny puzzle: Place eight pennies in a row. Transform them in four moves into four stacks of two coins each. On each move a single penny must jump exactly two pennies in either direction and land on the next single penny. Solution: Number the pennies from 1 to 8. Move 4 to 7, 6 to 2, 1 to 3, 5 to 8.

For similar recreations refer to Geometric Problems and Puzzles.

BIBLIOGRAPHY

Brooke, Maxey. *Fun for the Money: Puzzles and Games with Coins.* New York: Charles Scribner's Sons, 1963.

Gardner, Martin. *Mathematical Carnival.* New York: Alfred A. Knopf, 1965.

———. *The Second "Scientific American" Book of Mathematical Puzzles and Diversions.* New York: Simon & Schuster, 1961.

Kinnard, Clark, ed. *Encyclopedia of Puzzles and Pastimes.* New York: Citadel Press, 1946.

**(THE) COMMONER'S DILEMMA.** See LOGICAL PROBLEMS AND PUZZLES.

**COMMON OR VULGAR FRACTIONS.** See NUMBER PATTERNS, TRICKS, AND CURIOSITIES.

**COMPLETE CHESS.** See FAIRY CHESS.

**COMPOSITE NUMBERS.** See NUMBER PATTERNS, TRICKS, AND CURIOSITIES.

**CONCEALMENT CIPHERS.** See CRYPTARITHMS, CRYPTOGRAPHY, SUBSTITUTION CIPHERS, TRANSPOSITION CIPHERS, AND CODE MACHINES.

**CONTINUOUS LINE PUZZLES.** See GEOMETRIC PROBLEMS AND PUZZLES.

**CONTINUUM HYPOTHESIS.** See PARADOXES OF THE INFINITE.

**COURIER CHESS or THE COURIER GAME.** See FAIRY CHESS.

**CROSS-CAP.** See THE MÖBIUS STRIP.

**CROSS OF CARDS.** See CARD TRICKS AND PUZZLES.

**CRYPT-ARITHMETIC (CRYPTARITHMETIC).** See CRYPTARITHMS, CRYPTOGRAPHY, CONCEALMENT CIPHERS, SUBSTITUTION CIPHERS, TRANSPOSITION CIPHERS, AND CODE MACHINES.

**CRYPTARITHMIE.** See CRYPTARITHMS, CRYPTOGRAPHY, CONCEALMENT CIPHERS, SUBSTITUTION CIPHERS, TRANSPOSITION CIPHERS, AND CODE MACHINES.

**CRYPTARITHMS** (also called crypt-arithmetic, cryptarithmie, alphametic, and arithmetical restoration), **CRYPTOGRAPHY, CONCEALMENT CIPHERS, SUBSTITUTION CIPHERS, TRANSPOSITION CIPHERS, AND CODE MACHINES** are all involved in creating or deciphering messages.

Cryptarithms are mathematical problems involving the replacement of the digits by letters or some other symbols:

A × BC = BC

A + B = C

B × C = F

F − C − B = A

In the above equations, letters of the alphabet have been substituted for numbers sequentially (e.g., A = 1, B = 2, C = 3, and so on).

The term "cryptarithmie" (translated "cryptarithmetic") was first applied to this form of mathematical word play by Minos (M. Vatriquant), who first used it in the May 1931 issue of the Belgian magazine of recreational mathematics, *Sphinx,* along with the following example:

ABC

DE

FEC

DEC

HGBC

This problem can be solved as follows:

1. Since D × A = D, and D × other numbers does not equal those numbers, A = 1.

2. D × C and E × C both end in C; 5 is the only single digit that can accomplish this; thus, C = 5.

3. For C to equal 5, both D and E must be odd numbers (the only other possibility is for C to equal 0, which is impossible, in which case D and E would be even). Since A = 1, D and E cannot equal 1. Since their partial products produce three-digit numbers, they cannot be 9. Thus, they must be 3 and 7 (though not necessarily in that order). Since the partial product of D × B is only one digit, while that of E × B is two digits, E must be larger than D. Thus, E = 7, D = 3.

4. Since D × B has a one-digit total, B must be less than 3 (i.e., 0 or 2). B cannot be 0, because 0 times anything would equal 0. B = 2.

5. The rest are easy to fill in by simply completing the multiplication: F = 8, G = 6, and H = 4.

6. The answer: 125 × 37 = 4,625.

Maurice Kraitchik, editor of *Sphinx* from 1931 to 1939, points out that, though Vatriquant named the activity, better examples can be found previous to 1932 in

*La Mathématique des Jeux.* Kraitchik includes two examples in his book, *Mathematical Recreations.* One of them involves long division in which most of the numbers are covered by chessmen as follows (here *x*'s are used in place of the chessmen):

```
          xx8xx
      ┌─────────
xx │ xxxxxxx
      xxx
      ─────
        xx
        xx
        ───
        xxx
        xxx
        ───
          1
```

The solution is as follows: Since the five-digit quotient forms only three products with the divisor, two of the quotient's five digits must be zeroes. These zeroes cannot be either the first or the last digits, since both obviously result in products. Thus, the second and fourth digits of the quotient must be zeroes.

The two-digit divisor, when multiplied by 8, results in a two-digit product, but when multiplied by the first digit in the quotient results in a three-digit product. Thus, it must be larger than 8, i.e., it must be 9.

Since both the first and final digits in the quotient produce three-digit products with the two-digit divisor, they must be nines. Thus, the quotient equals 90,809.

When the divisor is multiplied by 8 it forms a two-digit product; when multiplied by 9 a three-digit product is formed. It must therefore be 12 (i.e., 8 × 12 = 96; 9 × 12 = 108).

This type of cryptarithm is a form of arithmetic restoration, and Maxey Brooke, *150 Puzzles in Crypt-Arithmetic,* points out that crypt-arithmetic is really a descendent of arithmetical restoration, which was probably invented in India during the Middle Ages. Arithmetic restorations, as in the above, attempt to reconstruct equations where various digits have been left out, for example:

```
  x1xxx
     417
  ─────
9xxx057
```

(The product = 9,141,057)

A cryptarithm where the letters substituted for the numbers also make sense linguistically is called an alphametic (the term was coined by J. A. H. Hunter in 1955). One of the oldest and best known is:

```
  SEND        9567
+ MORE      + 1085
─────       ──────
MONEY       10652
```

An alphametic is also a rebus, i.e., a representation of a word or a phrase by symbols or pictures. A rebus, however, is not necessarily an alphametic. The following rebus, for example, is not an alphametic, even though it is a play on the operations of mathematics:

$$\frac{Paid}{He} = \text{He is underpaid}$$

For more discussion and additional examples of alphametics refer to Joseph S. Madachy, *Mathematics on Vacation.*

Cryptography (*crypt* from the Greek *kryptos* ["secret"] and *graphy* from the Greek *grapho* ["writing"]) is the writing and deciphering of messages in secret code. David Kahn, *The Codebreakers,* provides a detailed history of the activity, including discussion of such interesting episodes in code breaking as the solving of Linear B by Michael Ventris and Alice Kober and the attempt by such men as Ignatius Donnelly to prove that Francis Bacon wrote Shakespeare's plays.

A *code* is a system of arbitrarily assigned symbols, often secret, for letters, words, or phrases. According to Helen Fouché Gaines, *Cryptanalysis: A Study of Ciphers and Their Solution,* a *cipher* is a coded message, generally one in which each letter of the alphabet is represented by a different letter or symbol. The difference between a code and a cipher is that a code requires a code book or dictionary to decode and can only be successful if the symbols are applied in a haphazard manner, whereas a cipher is based on some form of logical or mathematical formula. Many people tend to use the two terms interchangeably.

There are three main types of ciphers: concealment, transposition, and substitution. In a concealment cipher the true letters are hidden or disguised and are meant to pass unnoticed, as in written charades. A standard form of written charades goes as follows: a clue is given for the first syllable of the word; a clue is given for the second syllable of the word; and a clue is given for the entire word:

Seaside = the word to be guessed

My first syllable is a large body of water. (sea)
My second syllable is neither the top, bottom, front, nor back. (side)
My whole is a stretch of land that may have a beach. (seaside)

Another form of written charades involves the breaking up of words, phrases, or sentences at different points to produce different meanings:

I ran over. John and Bill wished I hadn't.
I ran over John, and Bill wished I hadn't.
I ran over John and Bill. Wished I hadn't.

Albrecht Dürer used a form of concealment cipher to date his picture, *Melancolia:*

He includes a magic square (*see* Magic Squares) with the date concealed in the middle squares of the bottom row.

In a transposition cipher the true letters are rearranged by some key or pattern, i.e., mathematical formula. Here are a few examples:

The *twisted path cipher* is a transposition cipher involving a matrix:

1. Start with the message, e.g., "Hide until dawn," and count the number of letters, e.g., 13.

2. Add enough letters to make a total that is a multiple of 4 (e.g., add $\underline{x}$, $\underline{y}$, $\underline{z}$ to the original 13 to total 16).

3. Place the letters in a grid or matrix:

| H | I | D | E |
|---|---|---|---|
| U | N | T | I |
| L | D | A | W |
| N | X | Y | Z |

4. Set up a path through the grid (e.g., a tiller's path, so named because it follows the same path as a farmer's path through a field):

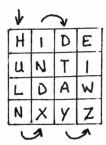

5. Line up the letters according to this path:

HULN XDNI DTAY EIWZ

The *rail fence cipher* is a transposition cipher which divides and mixes letters in groups of four, similar to the twisted path cipher:

1. Start with the message, e.g., "Shoot to kill," and count the number of letters, e.g., 11.

2. Add enough letters to total a multiple of 4 (e.g., add $\underline{h}$ to the 11 to total 12).

3. Write out the message in "rail fence" style:

$S_H O_O T_T O_K I_L L_H$

4. Copy the top row:

SOTOIL

5. Add the bottom row:

SOTOILHOTKLH

6. Divide up into groups of four:

SOTO ILHO TKLH

The *scrambling with a key word cipher* is a transposition cipher developed out of the twisted path cipher:

1. Start with the message, e.g., "Hide until dawn," and count the number of letters, e.g., 13.
2. Once again, add enough letters to make a total that is a multiple of 4 (e.g., x, y, z to the original 13 to total 16).
3. Place in a grid:

| H | I | D | E |
|---|---|---|---|
| U | N | T | I |
| L | D | A | W |
| N | X | Y | Z |

4. Choose a "key word" for mixing up the columns of the matrix, e.g., 2143, and match up with a word (or group of letters), e.g., T = 2, I = 1, M = 4, and E = 3.
5. Write this word that contains the key word (order of the columns) across the top of the grid:

T   I   M   E

| H | I | D | E |
|---|---|---|---|
| U | N | T | I |
| L | D | A | W |
| N | X | Y | Z |

6. Copy the four-letter columns out into four groups of letters according to the key word:

INDX HULN EIWZ DTAY

In a substitution cipher the order of the letters remains the same, but the letters or symbols are changed (substituted for). Here are a few examples:

The *Lewis Carroll Vigenère Cipher* is Lewis Carroll's version of a substitution code invented by Blaise de Vigenère, a sixteenth-century Frenchman. It goes as follows:

1. Start with a "key word" (e.g., Lewis Carroll used "vigilance"), and write it above the secret message as many times as necessary. Lewis Carroll chose "Meet me on Tuesday evening at seven" for his message:

   vigi la nc evigila ncevigi la ncevi
   Meet me on Tuesday evening at seven.

2. Set up the following alphabetic grid:

|   | A | B | C | D | E | F | G | H | I | J | K | L | M | N | O | P | Q | R | S | T | U | V | W | X | Y | Z |
|---|---|---|---|---|---|---|---|---|---|---|---|---|---|---|---|---|---|---|---|---|---|---|---|---|---|---|
| A | A | B | C | D | E | F | G | H | I | J | K | L | M | N | O | P | Q | R | S | T | U | V | W | X | Y | Z |
| B | B | C | D | E | F | G | H | I | J | K | L | M | N | O | P | Q | R | S | T | U | V | W | X | Y | Z | A |
| C | C | D | E | F | G | H | I | J | K | L | M | N | O | P | Q | R | S | T | U | V | W | X | Y | Z | A | B |
| D | D | E | F | G | H | I | J | K | L | M | N | O | P | Q | R | S | T | U | V | W | X | Y | Z | A | B | C |
| E | E | F | G | H | I | J | K | L | M | N | O | P | Q | R | S | T | U | V | W | X | Y | Z | A | B | C | D |
| F | F | G | H | I | J | K | L | M | N | O | P | Q | R | S | T | U | V | W | X | Y | Z | A | B | C | D | E |
| G | G | H | I | J | K | L | M | N | O | P | Q | R | S | T | U | V | W | X | Y | Z | A | B | C | D | E | F |
| H | H | I | J | K | L | M | N | O | P | Q | R | S | T | U | V | W | X | Y | Z | A | B | C | I | E | F | G |
| I | I | J | K | L | M | N | O | P | Q | R | S | T | U | V | W | X | Y | Z | A | B | C | D | E | F | G | H |
| J | J | K | L | M | N | O | P | Q | R | S | T | U | V | W | X | Y | Z | A | B | C | D | E | F | G | H | I |
| K | K | L | M | N | O | P | Q | R | S | T | U | V | W | X | Y | Z | A | B | C | D | E | F | G | H | I | J |
| L | L | M | N | O | P | Q | R | S | T | U | V | W | X | Y | Z | A | B | C | D | E | F | G | H | I | J | K |
| M | M | N | O | P | Q | R | S | T | U | V | W | X | Y | Z | A | B | C | D | E | F | G | H | I | J | K | L |
| N | N | O | P | Q | R | S | T | U | V | W | X | Y | Z | A | B | C | D | E | F | G | H | I | J | K | L | M |
| O | O | P | Q | R | S | T | U | V | W | X | Y | Z | A | B | C | D | E | F | G | H | I | J | K | L | M | N |
| P | P | Q | R | S | T | U | V | W | X | Y | Z | A | B | C | D | E | F | G | H | I | J | K | L | M | N | O |
| Q | Q | R | S | T | U | V | W | X | Y | Z | A | B | C | D | E | F | G | H | I | J | K | L | M | N | O | P |
| R | R | S | T | U | V | W | X | Y | Z | A | B | C | D | E | F | G | H | I | J | K | L | M | N | O | P | Q |
| S | S | T | U | V | W | X | Y | Z | A | B | C | D | E | F | G | H | I | J | K | L | M | N | O | P | Q | R |
| T | T | U | V | W | X | Y | Z | A | B | C | D | E | F | G | H | I | J | K | L | M | N | O | P | Q | R | S |
| U | U | V | W | X | Y | Z | A | B | C | D | E | F | G | H | I | J | K | L | M | N | O | P | Q | R | S | T |
| V | V | W | X | Y | Z | A | B | C | D | E | F | G | H | I | J | K | L | M | N | O | P | Q | R | S | T | U |
| W | W | X | Y | Z | A | B | C | D | E | F | G | H | I | J | K | L | M | N | O | P | Q | R | S | T | U | V |
| X | X | Y | Z | A | B | C | D | E | F | G | H | I | J | K | L | M | N | O | P | Q | R | S | T | U | V | W |
| Y | Y | Z | A | B | C | D | E | F | G | H | I | J | K | L | M | N | O | P | Q | R | S | T | U | V | W | X |
| Z | Z | A | B | C | D | E | F | G | H | I | J | K | L | M | N | O | P | Q | R | S | T | U | V | W | X | Y |

3. Since <u>M</u> is the first letter of the message and <u>V</u> is the first letter of the code word, find the column headed by <u>V</u> and move down it to the row designated by <u>M</u>. The letter in this square is <u>H</u>, which becomes the first letter of the coded message:

HMKB XEBP XPMY LLYR XIIQ TOLT FGZZ V

To decode the message, write the key word over it in the same manner that it was written over the original message, e.g., over the first letter <u>H</u> place a <u>V</u>. Then go to the column headed by <u>V</u>, move down it to <u>H</u>, and move across to the letter that designates that row, <u>M</u>.

A *shift cipher* (also called a *Caesar cipher,* because Julius Caesar used it) is a substitution cipher where each letter is substituted by a letter at some designated number of spaces deeper in the alphabet (e.g., if the number chosen is 5, then A = F, B = G, C = H, and so on).

A *date shift cipher* is a substitution cipher where a date (e.g., May 17, 1950; 5-17-50) is used to determine the shift of the letters. The dashes are removed (e.g., 51750) and the date is placed repeatedly over the message:

5175 05 17
Keep it up

The letters are then shifted deeper into the alphabet the number of spaces indicated by the number above each of them in turn (e.g., K = P, E = F, E = L, P = U, and so on).

A *key word cipher* is a substitution cipher where a key word is chosen and placed at the beginning of the alphabet to begin the shift of letters:

Money = key word

ABCDEFGHIJKLMNOPQRSTUVWXYZ
MONEYABCDFGHIJKLPQRSTUVWXZ

*The Shadow's code* is a substitution code where circles with lines in them are substituted for letters of the alphabet. The cipher was invented by Maxwell Grant (pseudonym of the creator of The Shadow, Walter B. Gibson). The Shadow was a 1930s mystery crime fighter in pulp magazines and on a popular radio show. He dressed all in black, and slipped through the darkness to fight evil. This code from *The Chain of Death* is but one of many he encountered:

$E = \otimes$    $P = \oplus$

$F = \oplus$    $Q = \ominus$

$G = \ominus$    $R = \ominus$     $1 = \ominus$

$H = \ominus$    $S = \otimes$     $2 = \ominus$

$I = \oslash$    $T = \otimes$     $3 = \oslash$

$J = \oslash$    $U = \otimes$     $4 = \ominus$

$K = \ominus$    $V = \otimes$

The symbols indicated by 1, 2, 3, 4 could be slipped into the message at any point to indicate which way to turn the paper in order to decipher the following circles.

The *pigpen cipher* (also called the Mason's cipher, because the Society of Freemasons used it some one hundred years ago) is a substitution cipher employing tictactoe and x patterns as follows:

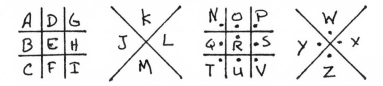

NEVER SAY DIE =

The *Polybius checkerboard cipher* is a substitution cipher named after its originator, an ancient Greek writer. The alphabet is written in a 5 × 5 grid (*y* and *z* occupying the same square):

|   | 1 | 2 | 3 | 4 | 5 |
|---|---|---|---|---|---|
| 1 | A | B | C | D | E |
| 2 | F | G | H | I | J |
| 3 | K | L | M | N | O |
| 4 | P | Q | R | S | T |
| 5 | U | V | W | X | Y/Z |

The digits indicating each square are then substituted for the letters (i.e., A = 11, B = 12, C = 13, and so on).

A *polyalphabetic cipher* is a substitution cipher where multiple symbols are used for the same letter or the same symbol is used for more than one letter (in the Polybius checkerboard cipher 55 stands for both $y$ and $\underline{z}$). Lewis Carroll's Vigenère cipher is a better example. Two additional examples follow:

*Porta's digraphic cipher* is a polyalphabetic substitution cipher where pairs (digraphic) of letters provide the basis for the text, i.e., a single symbol is substituted for every pair of letters in the message. It was invented by Giovanni Battista Porta, an Italian writer, scientist, and magician, and published in *The Code-breakers* (1563).

A giant 26 × 26 square grid is laid out as follows:

|   | A | B | C | D | E | F | G | H | I | J | K | L | M | N | O | P | Q | R | S | T | U | V | W | X | Y | Z |
|---|---|---|---|---|---|---|---|---|---|---|---|---|---|---|---|---|---|---|---|---|---|---|---|---|---|---|
| A | 1 | 2 | 3 | 4 | 5 | 6 | 7 | 8 | 9 | 10 | 11 | 12 | 13 | 14 | 15 | 16 | 17 | 18 | 19 | 20 | 21 | 22 | 23 | 24 | 25 | 26 |
| B | 27 | 28 | 29 | 30 | 31 | 32 | 33 | 34 | 35 | 36 | 37 | 38 | 39 | 40 | 41 | 42 | 43 | 44 | 45 | 46 | 47 | 48 | 49 | 50 | 51 | 52 |
| C | 53 | 54 | 55 | 56 | 57 | 58 | 59 | 60 | 61 | 62 | 63 | 64 | 65 | 66 | 67 | 68 | 69 | 70 | 71 | 72 | 73 | 74 | 75 | 76 | 77 | 78 |
| D | 79 | 80 | 81 | 82 | 83 | 84 | 85 | 86 | 87 | 88 | 89 | 90 | 91 | 92 | 93 | 94 | 95 | 96 | 97 | 98 | 99 | 100 | 101 | 102 | 103 | 104 |
| E | 105 | 106 | 107 | 108 | 109 | 110 | 111 | 112 | 113 | 114 | 115 | 116 | 117 | 118 | 119 | 120 | 121 | 122 | 123 | 124 | 125 | 126 | 127 | 128 | 129 | 130 |
| F | 131 | 132 | 133 | 134 | 135 | 136 | 137 | 158 | 139 | 140 | 141 | 142 | 143 | 144 | 145 | 146 | 147 | 148 | 149 | 150 | 151 | 152 | 153 | 154 | 155 | 156 |
| G | 157 | 158 | 159 | 160 | 161 | 162 | 163 | 164 | 165 | 166 | 167 | 168 | 169 | 170 | 171 | 172 | 173 | 174 | 175 | 176 | 177 | 178 | 179 | 180 | 181 | 182 |
| H | 183 | 184 | 185 | 186 | 187 | 188 | 189 | 190 | 191 | 192 | 193 | 194 | 195 | 196 | 197 | 198 | 199 | 200 | 201 | 202 | 203 | 204 | 205 | 206 | 207 | 208 |
| I | 209 | 210 | 211 | 212 | 213 | 214 | 215 | 216 | 217 | 218 | 219 | 220 | 221 | 222 | 223 | 224 | 225 | 226 | 227 | 228 | 229 | 230 | 231 | 232 | 233 | 234 |
| J | 235 | 236 | 237 | 238 | 239 | 240 | 241 | 242 | 243 | 244 | 245 | 246 | 247 | 248 | 249 | 250 | 251 | 252 | 253 | 254 | 255 | 256 | 257 | 258 | 259 | 260 |
| K | 261 | 262 | 263 | 264 | 265 | 266 | 267 | 268 | 269 | 270 | 271 | 272 | 273 | 274 | 275 | 276 | 277 | 278 | 279 | 280 | 281 | 282 | 283 | 284 | 285 | 286 |
| L | 287 | 288 | 289 | 290 | 291 | 292 | 293 | 294 | 295 | 296 | 297 | 298 | 299 | 300 | 301 | 302 | 303 | 304 | 305 | 306 | 307 | 308 | 309 | 310 | 311 | 312 |
| M | 313 | 314 | 315 | 316 | 317 | 318 | 319 | 320 | 321 | 322 | 323 | 324 | 325 | 326 | 327 | 328 | 329 | 330 | 331 | 332 | 333 | 334 | 335 | 336 | 337 | 338 |
| N | 339 | 340 | 341 | 342 | 343 | 344 | 345 | 346 | 347 | 348 | 349 | 350 | 351 | 352 | 353 | 354 | 355 | 356 | 357 | 358 | 359 | 360 | 361 | 362 | 363 | 364 |
| O | 365 | 366 | 367 | 368 | 369 | 370 | 371 | 372 | 373 | 374 | 375 | 376 | 377 | 378 | 379 | 380 | 381 | 382 | 383 | 384 | 385 | 386 | 387 | 388 | 389 | 390 |
| P | 391 | 392 | 393 | 394 | 395 | 396 | 397 | 398 | 399 | 400 | 401 | 402 | 403 | 404 | 405 | 406 | 407 | 408 | 409 | 410 | 411 | 412 | 413 | 414 | 415 | 416 |
| Q | 417 | 418 | 419 | 420 | 421 | 422 | 423 | 424 | 425 | 426 | 427 | 428 | 429 | 430 | 431 | 432 | 433 | 434 | 435 | 436 | 437 | 438 | 439 | 440 | 441 | 442 |
| R | 443 | 444 | 445 | 446 | 447 | 448 | 449 | 450 | 451 | 452 | 453 | 454 | 455 | 456 | 457 | 458 | 459 | 460 | 461 | 462 | 463 | 464 | 465 | 466 | 467 | 468 |
| S | 469 | 470 | 471 | 472 | 473 | 474 | 475 | 476 | 477 | 478 | 479 | 480 | 481 | 482 | 483 | 484 | 485 | 486 | 487 | 488 | 489 | 490 | 491 | 492 | 493 | 494 |
| T | 495 | 496 | 497 | 498 | 499 | 500 | 501 | 502 | 503 | 504 | 505 | 506 | 507 | 508 | 509 | 510 | 511 | 512 | 513 | 514 | 515 | 516 | 517 | 518 | 519 | 520 |
| U | 521 | 522 | 523 | 524 | 525 | 526 | 527 | 528 | 529 | 530 | 531 | 532 | 533 | 534 | 535 | 536 | 537 | 538 | 539 | 540 | 541 | 542 | 543 | 544 | 545 | 546 |
| V | 547 | 548 | 549 | 550 | 551 | 552 | 553 | 554 | 555 | 556 | 557 | 558 | 559 | 560 | 561 | 562 | 563 | 564 | 565 | 566 | 567 | 568 | 569 | 570 | 571 | 572 |
| W | 573 | 574 | 575 | 576 | 577 | 578 | 579 | 580 | 581 | 582 | 583 | 584 | 585 | 586 | 587 | 588 | 589 | 590 | 591 | 592 | 593 | 594 | 595 | 596 | 597 | 598 |
| X | 599 | 600 | 601 | 602 | 603 | 604 | 605 | 606 | 607 | 608 | 609 | 610 | 611 | 612 | 613 | 614 | 615 | 616 | 617 | 618 | 619 | 620 | 621 | 622 | 623 | 624 |
| Y | 625 | 626 | 627 | 628 | 629 | 630 | 631 | 632 | 633 | 634 | 635 | 636 | 637 | 638 | 639 | 640 | 641 | 642 | 643 | 644 | 645 | 646 | 647 | 648 | 649 | 650 |
| Z | 651 | 652 | 653 | 654 | 655 | 656 | 657 | 658 | 659 | 660 | 661 | 662 | 663 | 664 | 665 | 666 | 667 | 668 | 669 | 670 | 671 | 672 | 673 | 674 | 675 | 676 |

If the word "at" is to be put in code, the first letter, A, is found on the vertical column, and the second letter, T, is found on the horizontal column. The result is "20," which is then substituted for both letters, i.e., at = 20.

The *Playfair cipher* is another polyalphabetic substitution cipher using a grid. It was devised by Charles Wheatstone, a nineteenth-century British subject, friend of Baron Lyon Playfair, and used by the British army during the Boer War. The Australians used it in World War II. Dorothy L. Sayers' fictional detective Lord Peter Wimsey solves it in *Have His Carcase*.

A grid larger than twenty-six squares is laid out and filled randomly with letters (numbers other than 1, because 1 looks too much like I, are used to fill out the grid):

| I | J | H | T | 10 | 13 | 4 |
|---|---|---|----|----|----|---|
| A | K | S | 12 | C | D | 7 |
| 17 | R | B | Z | 14 | E | 9 |
| 11 | 16 | Y | 15 | Q | V | U |
| W | 3 | 18 | 19 | F | G | P |
| 5 | L | X | 20 | 21 | 6 | O |
| 2 | M | 22 | 24 | 23 | 8 | N |

The rules are as follows:

1. The message is encoded by taking a pair of letters at a time.

2. If both letters in a pair are in the same row on the grid, the letters immediately to the right of each letter are used. Each row is thought of as being attached to the row above it horizontally (e.g., *A* comes right after *4*).

3. If both letters are in the same column, the letters immediately below are used. In this case the columns are thought of as being attached to the previous column vertically (e.g., *J* comes right after *2*).

4. If the letters are in different rows and different columns, the letters are replaced, each letter in turn, by the letter or number in the same row that is also in the column occupied by the other letter, e.g., RI; R = 17, I = J; RI = 17J.

5. If, when the letters of a message are divided into twos, the same letter is repeated, an *x* is inserted to divide the two letters. If there is a letter left over, an *x* is added to make it a pair.

A *random substitution cipher,* similar to the one above, is a substitution cipher where the substitution symbols are assigned at random and a key (an index, a code book) is necessary. Examples of these can be found in "The Adventure of the Dancing Men" by Arthur Conan Doyle, and "The Gold Bug" by Edgar Allan Poe.

*Code machines* are machines used to create and decipher codes. Here are a few examples:

*Grilles* are grids with holes cut in them to create "windows" through which the message can be written or deciphered. After the message has been written in, the rest of the spaces are filled in with random numbers. Jules Verne includes the use of a grille in *Mathias Sandorff* (1885).

*Thomas Jefferson's wheel cipher machine* consists of thirty-six wooden wheels of the same size mounted on an iron rod, each wheel with a scrambled alphabet around its edge. The wheels are aligned so that they spell out a message. Then the letters from a different place on the wheels are copied down. This is continued throughout the message.

The *scytale* is the Latin name of the earliest known code device. It was first used by the Spartans in 500 B.C. A strip is wrapped around a cylinder and a message written on it. When the strip is unwrapped, the message is mixed.

The *Alberti disk* is a substitution code machine where one disk is placed inside another so that it can be rotated. The letters of the alphabet are placed around the circumference of each disk. By turning the inside disk, the letters can be lined up on the inside disk with other letters on the outside disk and substituted for them.

*Typewriter codes* are created by using relationships of letters on a typewriter keyboard to determine substitutions. *Telephone dial codes* are created by using the letter combinations on the dial of a telephone.

For similar mathematical play refer to Number Patterns, Tricks, and Curiosities.

BIBLIOGRAPHY

Ball, W. W. Rouse. *Mathematical Recreations and Essays.* 1892. Rev. by H. S. M. Coxeter. London: Macmillan and Co., 1939.
Brandreth, Gyles. *Indoor Games.* London: Hodder & Stoughton, 1977.
————. *The World's Best Indoor Games.* New York: Pantheon Books, 1981.
Brooke, Maxey. *150 Puzzles in Crypt-Arithmetic.* 2d ed., rev. New York: Dover, 1969.
Gaines, Helen Fouché. *Cryptanalysis: A Study of Ciphers and Their Solution.* New York: Dover, 1939.
Gardner, Martin. *Codes, Ciphers and Secret Writing.* New York: Simon and Schuster, 1972.
Kahn, David. *The Codebreakers: The Story of Secret Writing.* New York: Macmillan and Co., 1968.
Kinnard, Clark, ed. *Encyclopedia of Puzzles and Pastimes.* New York: Citadel Press, 1946.

Madachy, Joseph S. *Mathematics on Vacation*. New York: Charles Scribner's Sons, 1966.

Schaaf, W. L. "Number Games and Other Mathematical Recreations." In *Macropaedia: Encyclopedia Britannica*, 1985.

Shipley, Joseph T. *Playing with Words*. Englewood Cliffs, N.J.: Prentice-Hall, 1960.

**CRYPTOGRAPHY.** See CRYPTARITHMS, CONCEALMENT CIPHERS, SUBSTITUTION CIPHERS, TRANSPOSITION CIPHERS, AND CODE MACHINES.

**CUBE-O.** See COIN PROBLEMS AND PUZZLES.

**CYCLIC NUMBERS.** See NUMBER PATTERNS, TRICKS, AND CURIOSITIES.

**CYLINDRICAL CHESS.** See FAIRY CHESS.

**D**

**DATE SHIFT CIPHER.** See CRYPTARITHMS, CRYPTOGRAPHY, CONCEALMENT CIPHERS, SUBSTITUTION CIPHERS, TRANSPOSITION CIPHERS, AND CODE MACHINES.

**DELIAN PROBLEM.** See THE DUPLICATION OF THE CUBE.

**DIABLOTIN.** See THE FIFTEEN PUZZLE.

**DIABOLIC SQUARES.** See MAGIC SQUARES.

**DICE TRICKS** are number tricks involving the use of dice.

The following trick is included in both Martin Gardner's *Mathematics, Magic and Mystery* and Boris A. Kordemsky's *The Moscow Puzzles:*

A magician turns his back while a spectator throws three dice on a table. The dice thrower is then instructed to add the faces. He is then asked to pick up any one die and add the number on the bottom of it to the previous total. This same die is rolled again and the number shown added to the total.

The magician turns around, calls attention to the fact that he has no way of knowing which of the three dice was used for the second roll, picks up the dice, shakes them for a minute, and announces the total.

The solution: Before the magician picks up the dice, he totals their faces. Seven added to this number gives the total obtained by the spectator.

Kordemsky includes the following similar trick: Whoever is performing the trick turns his back and has someone roll three dice and arrange them so they form a three-digit number, say, 254. Then the person who rolled the dice is asked to arrange the three numbers on the bottom of the dice on the right side of the original three-digit number, e.g., 254,523. Finally, this person is asked to divide this number by 111 and tell the person performing the trick the result, e.g., 2,293. The person performing the trick can then tell what the top faces of the original three dice were.

The solution: Seven is subtracted from the number given, and then the number is divided by nine, e.g., 2,293 − 7 = 2,286; 2,286/9 = 254.

The solutions to these tricks have to do with the fact that opposite sides of a die always add up to seven. For a more lengthy discussion of this type of problem and the mathematics involved in solving it refer to Number Patterns, Tricks, and Curiosities.

BIBLIOGRAPHY

Collins, A. Frederick. *Fun with Figures*. New York: D. Appleton and Co., 1928.
Gardner, Martin. *Mathematics, Magic, and Mystery*. New York: Dover, 1956.
Kordemsky, Boris A. *The Moscow Puzzles*. Trans. by Albert Parry; ed. by Martin Gardner. New York: Charles Scribner's Sons, 1972.

**THE DICHOTOMY.** See ZENO'S PARADOXES.

**DIFFICULT CROSSINGS.** See LOGICAL PROBLEMS AND PARA-DOXES.

**DIGITAL PROBLEMS** are arithmetic and algebraic tasks involving curious properties of digits.

For example, write a number, say, 9, using all of the digits once and only once (the operations of arithmetic are allowed):

| 97,524, | 95,823, | 95,742, | 75,249, | 58,239, | 57,429 |
|---------|---------|---------|---------|---------|--------|
| 10,836  | 10,647  | 10,638  | 08,361  | 06,471  | 06,381 |

A common problem is to write 100 using all of the digits once and only once:

$$1 + 93 + 5 + \frac{4}{28} + \frac{6}{7}$$

or

$$97 + \frac{5+3}{8} + \frac{6}{4} + \frac{1}{2}$$

or

$$57 + 42 + \frac{9}{18} + \frac{3}{6}$$

or

$$1 + 2 + 3 + 4 + 5 + 6 + 7 + (8 \times 9).$$

There are numerous solutions to the above problems, and additional restrictions may be added to make the task more difficult. For example, use each of the digits in order once and only once to equal 100:

$$123 - 45 - 67 + 89 = 100$$

Once again, numerous solutions are possible, as is the case when the task is to use all the digits in reverse order:

$$98 - 76 + 54 + 3 + 21 = 100$$

The *problem of the four n's* requires the writing of the sequence of digits in order beginning with 1 by four of a selected digit used only four times (the rules of operations allowed may be varied). In the following, the factoral (!) is allowed:

$$1 = 1 + \frac{1}{1} - 1$$

$$2 = 1 + 1 + 1 - 1$$

$$3 = 1 + 1 + \frac{1}{1}$$

$$4 = 1 + 1 + 1 + 1$$

$$5 = (1 + 1 + 1)! - 1$$

and so on . . .

The four digits may be some other digit than 1:

$$1 = \frac{2}{2} + 2 - 2$$

$$2 = \frac{2}{2} + \frac{2}{2}$$

$$3 = (2 \times 2) - \frac{2}{2}$$

$$4 = 2 + 2 + 2 - 2$$

$$5 = (2 \times 2) + \frac{2}{2}$$

and so on . . .

A variation may require the use of the first four digits once each:

$$1 = (3 - 1)\frac{2}{4}$$

$$2 = 4 - 3 + 2 - 1$$

$$3 = \frac{2}{1} + 4 - 3$$

$$4 = 4 + 3 - 2 - 1$$

$$5 = 4 + 3 - \frac{2}{1}$$

and so on . . .


Joseph S. Madachy, *Mathematics on Vacation,* includes Henry E. Dudeney's solutions to the smallest square number containing the nine different digits (excluding zero):

$139,854,267 \ (11,826)^2$

and the largest:

$923,187,456 \ (30,384)^2$

For similar problems refer to Number Patterns, Tricks, and Curiosities.

BIBLIOGRAPHY

Bakst, Aaron. *Mathematical Puzzles and Pastimes*. 2d ed. Princeton, N.J.: D. Van Nostrand Co., 1965.

———. *Mathematics: Its Magic and Mastery*. New York: D. Van Nostrand Co., 1941.

Beiler, Albert H. *Recreations in the Theory of Numbers: The Queen of Mathematics Entertains*. 2d ed. New York: Dover, 1966.

Collins, A. Frederick. *Fun with Figures*. New York: D. Appleton and Co., 1928.

Kraitchik, Maurice. *Mathematical Recreations*. 2d rev. ed. New York: Dover, 1953.

Madachy, Joseph S. *Mathematics on Vacation*. New York: Charles Scribner's Sons, 1966.

Schaaf, W. L. "Number Games and Other Mathematical Recreations." In *Macropaedia: Encyclopedia Britannica,* 1985.

**DIGITAL ROOTS.** See GUESSING NUMBERS.


**DIOPHANTINE EQUATIONS.** See NUMBER PATTERNS, TRICKS, AND CURIOSITIES.


**DISSECTION PUZZLES.** See GEOMETRIC DISSECTIONS.


**DIVIDING THE PLANE PUZZLES.** See GEOMETRIC PROBLEMS AND PUZZLES.

**DOMINOES** (including numerous variations on the standard game and aspects of polyominoes) is a game based on a set of twenty-eight rectangular pieces of wood (or some other material) whose top surfaces are divided in half, with zero to six dots placed on each half, e.g.:

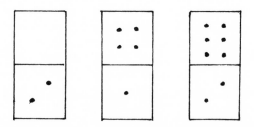

The standard game varies, but here are some general rules. The dominoes are mixed face down. The first player is determined by drawing dominoes (highest domino goes first, or, in the case of a tie, the domino with the highest single half, e.g., a domino with a 6-1 goes before a domino with a 4-3). Each player then draws a certain number of dominoes from the stock (pot, pool); if two or three players are competing, each usually draws seven or nine dominoes; if four, seven; if five, five. The remaining dominoes are then eliminated from the game or used as a pool to draw from when a player is unable to play (depending on the rules established).

The first player plays a domino (rules vary as to whether a certain domino must be played, e.g., his highest double). In one version, he continues to play as long as he can keep matching one of the ends of one of his dominoes with one of the ends of the dominoes on the table.

In another version, the play passes after a single domino has been played. The second player then matches one of the ends of one of his dominoes with one of the open ends of the domino on the table (if he cannot play, he either passes, or draws one until he can play, according to the rules established).

A domino with an equal number of pips (dots) on both ends (a double) may be or must be played crosswise. Depending on the rules, this may or may not create four ends to be played off of, e.g.:

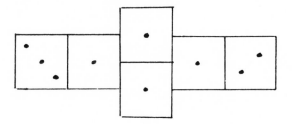

Play continues until one player uses up all of his dominoes. If no player is able to play any more dominoes and all of the players still have dominoes remaining, then the players reveal their remaining dominoes, the pips are counted up, and the player with the fewest pips wins. If a number of games are going to be played, sometimes a total score is established, say, 100, a certain number of points is awarded to the winner of each game, say, 5, plus the number of pips remaining in the opponents' hands, and a running total is kept.

Not only are there a number of slight variations in the standard game of dominoes, but there are also a number of variations that have taken on the status of games separate from the standard game. Here are some of these:

*Muggins* challenges a player to achieve the highest score by placing dominoes on the table so that the pips at the ends of the line add up to 5 (or a multiple of 5) in addition to being the first player to use all of his dominoes. Generally, doublets must be placed crosswise (no play allowed from their ends) and all of the pips on both ends counted to add up the score.

A player scores 1 point every time he makes pips add up to 5 or a multiple of 5 (1 point for 5; 2 points for 10; 3 points for 15; etc.). The first player to use up all of his dominoes scores 1 point for every 5 pips left on his opponents' dominoes (rounded to lowest 5). Play continues until someone reaches at least 61 points.

Before beginning play the dominoes are placed face down and mixed up. If two players are competing, each draws seven dominoes; if three, six; if four, five. The remaining dominoes are placed in a pool.

The player with the highest double plays first (if a double 5, he scores 2 points). The following players must play dominoes that match those in play. If a play is possible it must be made. If a player cannot play, he draws from the pool until he gets a domino he can play. If the pool is exhausted, he passes until he can play. When one of the players has used all of his dominoes, the round is over and the dominoes are remixed.

*Five Up* is a slight variation on Muggins. Each player begins with five dominoes, regardless of how many are playing. The first play is made by a player chosen by lot (instead of with the highest double), who may play any domino he wishes.

*All Fives* is a slight variation of Five Up. Whenever a player scores a point (or more) on a turn, he continues play (drawing dominoes, if necessary); and a player cannot go out with a domino scoring points (if that is all he has left, he must draw; if all the dominoes are drawn, he must pass).

*Sniff* is another variation on Muggins. The first player is chosen by lot and plays any domino he wishes. The major difference between Sniff and Muggins is in the use of the double. In Sniff, the first double in a round is called sniff and can be played either with or across the line. If sniff is played across the line of play, the open side must be played on before the other sides are played on; if played across the line, the sides may be played on, but only after the open end has been played on. All doubles after sniff must be played across the line of play.

*Doubles* begins with the standard mixing of the dominoes face down. If three, four, or five players are competing, each draws seven dominoes. The remaining dominoes form the pool.

The first player, chosen by lot, must play a double (if he does not have one, he must draw one). The second player must match the double with one end of one of his dominoes or draw until he can. The next player must either match the unused side of the double with a single end, or match the open single end with a double (or draw until he can). A pattern is thus created of alternating singles and doubles. This continues until one player gets rid of all of his dominoes.

*Ends* requires four players. The dominoes are mixed face down and each player draws seven (the players are not allowed to see each other's hands). The player with the double six begins play. Players take turns matching the dominoes. Each player can play only one domino in a turn.

If a player cannot play, he then asks the player to his left for a domino. That player gives him a domino if he has one and then takes his own turn. If the player to the left cannot play, the process continues until someone can play. If no one can play, the first player who could not play plays any domino, and play continues. The first player to use all of his dominoes wins.

*Round-the-Clock* requires, first, the choosing of a leader. Then, once again, the dominoes are mixed face down and drawn: if two players, seven; if three, six; if four or five, five. The remaining dominoes are placed in a pool.

The player with a double six begins play. If no one has a double six, the leader draws from the pool. If he draws the double six, he plays it; if not, the player on his left draws. This continues until someone draws it or there are only two dominoes left in the pool, in which case the game is abandoned.

The player to the left then plays, matching an end of one of his dominoes on either side or end of the double six. If he cannot play and there are more than two dominoes remaining in the pool, he draws one. If he still cannot play, he passes.

This continues until four dominoes have been played against the double six. Then four doubles must be played against the opposite ends of the four dominoes played against the double six. This pattern continues until someone uses up his dominoes or no one can play—in which case the player with the fewest remaining pips wins.

*Maltese Cross* is the same as Round-the-Clock, except that it is not necessary to wait until one row is finished before beginning the next. It gets its name from the shape created by the dominoes.

*Matador* again requires that the dominoes be mixed face down and drawn: if two players, seven each; if three, six; if four, five. The remaining dominoes are put in a pot.

The first player places any domino on the table. The second player must play a domino so that the ends placed together total seven:

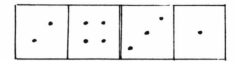

Or he can play a "matador" (a domino totaling seven all by itself or a double blank). Matadors are played across the line. The next domino must either total seven or be another matador (this time played with the line):

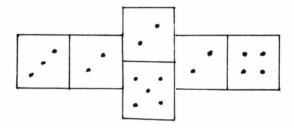

A player may decide not to use a matador and draw instead, in which case he continues to draw until he has another play or the pile is used up and he passes. When one player uses up his dominoes or all players pass in succession, the game ends. Players score 1 point for every join and 1 for every matador. The player using up his dominoes also scores 1 point for every remaining pip in the opponents' hands. If no one goes out, the players lose 1 point for every pip in their own hand and score 1 point for every pip in the opponents' hands. Play continues until someone reaches 101 points.

*French Draw* begins by choosing the first player by lot, mixing up the dominoes face down, and drawing: if two or three players, seven; if four, six. The remaining dominoes form a pool.

The first player plays any domino he wishes. The second player matches either end of the first domino or draws dominoes from the pool (as long as at least two are left undrawn) until he is able to play. Play continues until someone goes out or no one can play. The winner is the player who goes out or has the fewest pips.

*Block* begins by choosing the first player by lot and drawing dominoes: if two players, seven; if three, six; if four, five. The remaining dominoes are placed in a pool.

The first player plays any domino he wishes. The second player matches one end of the first domino with one of his own or draws until he can or until all of the dominoes are used, in which case he passes.

Doubles are played across the line, but only sides can be played on. Play continues until one player plays all of his dominoes or until no one can play. Each player receives 1 point for every pip remaining in his opponents' hands. Play continues until a player reaches 101 points.

*Bergen* is a variation of Block. A player is chosen to go first, and dominoes are drawn as in Block. The first player plays any domino he wishes (a double scores 2 points). If successive players can make one end of a line match the opposite end of the line, they score 2 points. If the match is made with a double domino, 3 points are scored. A match made off of a double also scores 3 points.

When someone plays all of his dominoes or no one can play, the player with the greatest number of points in that hand gets 1 extra point for each domino remaining in his opponents' hands, and loses 1 point for each remaining in his own hand.

*Tiddly-Wink* is identical to Block, except that whenever a player plays a double he is allowed to play a second domino if he wishes.

*Draw or Pass* begins by choosing the first player by lot, mixing the dominoes face down, and drawing: if two or three players, seven each; if four, six. The remaining dominoes go into a pool.

The first player plays any domino he wishes. The second player matches one end, draws as many dominoes from the pool as he wishes (so long as at least two remain), or passes. A player may not draw and play in the same turn. A player may pass, if he wishes, even though he could play. Play continues until one player goes out or all players pass in succession (in which case the player with the fewest pips wins).

*Sebastopol* is the same as Block, except for the following: It must have either three or four players. If there are four players, each player draws seven dominoes, and the player with the double six plays that. If there are three players, each player draws nine dominoes, and the first player is chosen by lot. If someone draws the double six it is played immediately and replaced with another draw. Then the chosen player begins play. If the double six remains in the pool after the draw, it is simply played first and then the lead player begins.

The first four plays must be to the double six (a domino on each end and on each side). All other doubles are played as if they were no different than the non-doubles.

Dominoes have also been used by mathematicians for recreations other than the standard game and its variations. The following is a magic square (*see* Magic Squares) where the pips on each domino added up produce an equal number for all rows, columns, and diagonals of 33:

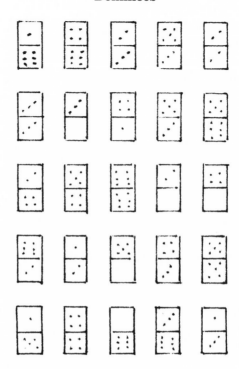

*Polyominoes* are dominoes of *n* squares, e.g.:

monomino = single square
domino     = two connected squares
tromino    = three connected squares
tetromino  = four connected squares
pentomino = five connected squares
hexomino  = six connected squares
and so on . . .

Henry Ernest Dudeney published the first pentomino problem in *Canterbury Puzzles* (1907), and in the 1930s and 1940s, the *Fairy Chess Review,* a British puzzle journal, published extensive literature on pentominoes. However, the beginnings of exploration of pentominoes go back to ancient Japan, when a master of the Japanese game of go realized that there are only twelve possible patterns for five connected squares.

The interest in pentominoes, nevertheless, at least as a form of combinatorial geometry, really began in 1954, when Solomon W. Golomb published an article on them in *American Mathematical Monthly.* Piet Hein coined the term ''polyomino'' and placed pentominoes within this class.

There are only nine possible shapes (assuming rotation and reflection are identical) of polyominoes of fewer than five squares:

MONOMINO

DOMINO

STRAIGHT TROMINO

RIGHT TROMINO

STRAIGT TETROMINO

SQUARE TETROMINO

T· TETROMINO

SKEW TETROMINO

L TETROMINO

Though the domino has been the standard polyomino for popular game play, modern mathematicians have found that pentominoes are far better suited for puzzle and game play. There are twelve distinct patterns five squares can be joined in to form a pentomino. These twelve patterns, then, make up a set of pentominoes:

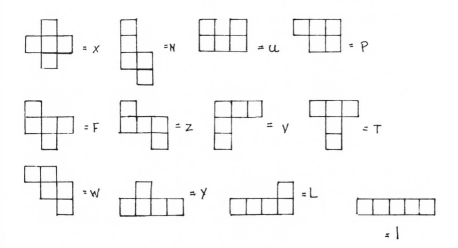

The object in pentomino problems is to construct rectangles of various sizes using all twelve pieces (e.g., construct a rectangle that is $5 \times 12$ or $6 \times 10$ or $4 \times 25$ or $3 \times 20$).

In a *triplication problem,* the object is to choose one of the pieces and use nine of the remaining eleven to construct a large-scale (three times the height and three times the width) replica of the chosen piece.

A standard game of pentominoes (called Golomb's Game after Solomon W. Golomb) goes as follows: The twelve pentominoes are cut out so that their squares are equal to the squares of a checkerboard. The two players sit across from each other, as if they were going to play checkers. The first player chooses any pentomino and places it on the board to cover five squares. The second player chooses another piece and covers any five spaces not yet covered. Play continues until one player is unable to play and loses.

For similar play refer to Geometric Dissections; Graphs and Networks; Tangrams.

BIBLIOGRAPHY

Brandreth, Gyles. *Indoor Games.* London: Hodder & Stoughton, 1977.
Collins, A. Frederick. *Fun with Figures.* New York: D. Appleton and Co., 1928.
Dudeney, Henry Ernest. *The Canterbury Puzzles.* 1919. Reprint. New York: Dover, 1958.
————. *536 Puzzles and Curious Problems.* Reprint Ed. by Martin Gardner. New York: Dover, 1967.
Golomb, Solomon W. *Polyominoes.* New York: Charles Scribner's Sons, 1965.
Kordemsky, Boris A. *The Moscow Puzzles.* Ed. by Martin Gardner; trans. by Albert Parry. New York: Charles Scribner's Sons, 1972.
Kraitchik, Maurice. *Mathematical Recreations.* 2d rev. ed. New York: Dover, 1953.
Schaaf, W. L. "Number Games and Other Mathematical Recreations." In *Macropaedia: Encyclopedia Britannica,* 1985.

**DOUBLE CENTURION.** See CENTURION.

**DOUBLES.** See DOMINOES.

**DRAUGHTS.** See CHECKERS.

**DRAW OR PASS.** See DOMINOES.

**THE DUPLICATION OF THE CUBE,** also called the Delian problem, is, along with the trisection of an angle and squaring the circle, one of the three famous problems of classic geometry.

The task is as follows. Find the edge of a cube that is double that of a given cube, using nothing more than a straight-edge and compasses to construct a finite number of straight lines and circles.

According to W. W. Rouse Ball, *A Short Account of the History of Mathematics* and *Mathematical Recreations and Essays,* the problem was originally known as the Delian problem, a result of a legend that the Delians had consulted Plato about it. According to Philoponus, the Athenians, suffering from a typhoid fever plague, consulted the oracle at Delos in 430 B.C. In reply, Apollo told them to double the size of his altar, which was in the form of a cube. They immediately constructed a new altar with edges twice that of the old one, thus increasing its volume eightfold. The indignant god created an even worse plague and told the next deputation that his new altar must be a cube exactly double in volume to that of the old one. The Athenians asked Plato what to do, and he referred them to Euclid (Plato and Euclid are obvious anachronisms).

James H. Weaver, "The Duplication Problem," includes another version of the story from Eutocius (c. 480 B.C.) preserved in a letter written by Eratosthenes (276–194 B.C.) to King Ptolemy (Archimedes, *Opera Omnia cum Commentariis Eutocii,* ed. by J. L. Heiberg. Leipzig, 1880–1881, Vol. 3, pp. 102–107). It goes as follows:

It is related that one of the old tragic poets whom Minos had imported says that when Minos wished to erect for his son Glaucus a tomb and noticed that its dimensions were one hundred feet

$$\acute{\epsilon}\kappa\alpha\tau o\mu\ \pi\epsilon\delta o s$$

on all sides he exclaimed: "You have enclosed too small a space for a royal tomb. Double it, but forget not the beautiful form. Therefore double each edge of the monument."

But he had clearly erred. For by doubling the edges, the surface is made four times as great and the volume is increased eight fold. Nevertheless it made the question of how a body could be doubled without changing its form an object of investigation among geometers, and this is called the duplication of the cube. That is, they set up a given cube and sought to double it.

The following Arab version of the legend is included in Ball's *Mathematical Recreations and Essays:*

Now in the days of Plato a plague broke out among the children of Israel. Then came a voice from heaven to one of their prophets, saying, "Let the size of the cubic altar be doubled, and the plague will cease"; so the people made another altar like unto the former, and laid the same by its side. Nevertheless the pestilence continued to increase. And again the voice spake unto the prophet, saying, "They have made a second altar like unto the former, and laid it by its side, but that does not produce the duplication of the cube.'' Then applied they to Plato, the Grecian sage, who spake to them, saying, ''Ye have been neglectful of the science of geometry, and therefore hath God chastised you, since geometry is the most sublime of all the sciences.'' Now, the duplication of a cube depends on a rare problem in geometry, namely . . .

Hippocrates of Chios (c. 420 B.C.) reduced the problem to one of finding two mean proportionals to two given lines, e.g., if two lines $a$ and $2a$ are given and if two mean proportionals can be inserted between them such that $a:x = x:y = y:2a$, then $x^2 = ay$ and $y^2 = 2ax$, which results in $x^3 = 2a^3$. Hippocrates, however, could not find the mean proportionals.

Archytus of Tarentum (c. 400 B.C.) accomplished this through the intersections of solids. Let $AD$ be the larger of the two given lines. Then on $AD$ as a diameter draw a circle and, using this circle as a base, construct a cylinder. From $A$ in the circle, draw chord $AB$ equal to the smaller of the two given lines and extend this chord to the tangent drawn from point $D$. Designate this as point $P$. Let the triangle $APD$ revolve around $AD$ as an axis generating a cone. Suppose a semicircle drawn on $AD$ as diameter and perpendicular to the circle $ADB$, and let this semicircle be moved so that $A$ remains fixed and the semicircle remains perpendicular to the circle $ABD$. It will form a solid. The intersection of the three solids will determine a point. From this point draw an element of the cylinder, and let this cut the circle $ABD$ in $C$. Draw $AC$. From this it can be shown that $AC^3 = 2AB^3$.

Menaechmus (c. 340 B.C.) came up with conic sections and used them to give two solutions to the problem. These consisted of (1) finding the intersection of two parabolas ($x^2 = ay$ and $y^2 = 2ax$) other than the one at the origin; and (2) the intersection of the hyperbola $xy = 2a^2$ and the parabola $x^2 = ay$.

Plato (c. 340 B.C.) offered a mechanical solution. Nicomedes solved it through the use of the conchoid, and Diocles solved it through the use of the cissoid. Nevertheless, in all of these solutions, the rules of the game (use of only straightedge and compasses) had been broken.

In 1593 François Viète (*Opera Mathematica*, ed. by Van Schooten. Leyden, 1646, prop. v, pp. 242–243), showed that every cubic or biquadratic that is not otherwise reducible leads to either the duplication or the trisection problem when solved. His construction is as follows: Describe a circle with the center $O$ and radius equal to half the length of the larger of two given lines. Draw chord $AD$ equal to the smaller of the two lines. Draw $AB$ to $E$ so that $BE = AB$. Draw line $AF$ parallel to $OE$ through $A$. Draw line $DOCFG$ through $O$, intersecting the circumference at $D$ and $C$, cutting $AF$ at $F$ and $BA$ at $G$, so that $GF = OA$. If this line is possible, then $AB:GC = GC:CA = GA:CD$.

In 1637 René Descartes (*Geometria*, ed. by Von Schooten. 1659, bk. iii, p. 91), after showing (in book ii) that plane problems corresponded to equations of the first and second degree, solid problems to equations of the third degree, and linear problems to equations of the fourth degree, used a circle to give a solution equivalent to that of Menaechmus. Descartes' solution goes as follows: curves $x^2 = ay$ and $x^2 + y^2 = ay + bx$ cut in a point $(x,y)$ such that $a:x = x:y = b$.

Gregoire de Saint-Vincent (*Opus Geometricum Quadraturae Circuli*. Antwerp, 1647, bk. v, prop. 138, p. 602) came up with a construction based on the following theorem: Curves $xy = ab$ and $x^2 + y^2 = ay + bx$ cut in a point $(x,y)$ such that $a:x = x:y = y:b$.

Contemporary mathematics (analytical geometry) is able to state absolutely (a dangerous word) that the duplication of a cube, along with the other two famous construction problems (the trisection of an angle and the squaring of a circle), cannot be done through the use of a straight-edge and compasses employed a limited number of times. This is based on the following theorem:

The necessary and sufficient condition that an analytic expression can be constructed with straight-edge and compasses is that it can be derived from the known quantities by a finite number of rational operations and square roots. For a detailed discussion of this refer to Felix Klein, *Famous Problems of Elementary Geometry*.

The equation for the duplication of a cube is: $x^3 = 2$, an irreducible equation of the third degree. Plane problems (those of the first and second degrees) can be solved by straight-edge and compasses; solid problems (those of the third degree) require conic sections; and linear problems (those of higher degrees) cannot be solved by either straight lines and circles or conics.

For similar mathematical recreations refer to Euclid's Theory of Parallels; Squaring the Circle; The Trisection of an Angle.

BIBLIOGRAPHY

Ball, W. W. Rouse. *Mathematical Recreations and Essays*. 1892. Rev. by H. S. M. Coxeter. London: Macmillan and Co., 1939.

———. *A Short Account of the History of Mathematics*. 1888. Reprint. London: Macmillan and Co., 1935.

Bell, E. T. *The Development of Mathematics*. New York: McGraw-Hill, 1945.

Cadwell, J. H. *Topics in Recreational Mathematics*. London: Cambridge University Press, 1966.

Carslaw, H. S. "On the Constructions Which Are Possible by Euclid's Methods." *Mathematical Gazette* 5 (1910): 171–179.

Clark, M. E. "Construction with Limited Means." *American Mathematical Monthly* 48 (1941): 475–479.

Dickson, L. E. "Constructions with Ruler and Compasses: Regular Polygons." In *Monographs on Topics of Modern Mathematics Relevant to the Elementary Field*. New York: Dover, 1955.

Gow, James. *A Short History of Greek Mathematics*. London: Cambridge University Press, 1884.

Klein, Felix. *Famous Problems of Elementary Geometry: The Duplication of the Cube, the Trisection of an Angle, the Quadrature of the Circle*. Trans. by Wooster Woodruff Beman and David Eugene Smith. New York: Chelsea Publishing Co., 1955; 2d ed., 1962.

Weaver, James H. "The Duplication Problem." *American Mathematical Monthly* 23 (1916): 106–113.

# E

**EGYPTIAN CALENDAR.** See CALENDARS.

**ENDGAME.** See CHESS PROBLEMS.

**ENDS.** See DOMINOES.

**EUCLID'S FIFTH POSTULATE.** See EUCLID'S THEORY OF PARALLELS.

**EUCLID'S PARALLEL POSTULATE.** See EUCLID'S THEORY OF PARALLELS.

**EUCLID'S THEORY OF PARALLELS,** also called Euclid's Fifth Postulate, Euclid's Parallel Postulate, the Postulate of Parallels, and the Fifth Postulate, challenges the mathematician to prove that only one line can be drawn through a given point parallel to a given line, and in so doing to prove Euclid's Fifth Postulate (translated by Sir Thomas L. Heath from the critical edition of J. L. Heiberg's *The Thirteen Books of Euclid's Elements)*:

If a straight line falling on two straight lines make [*sic*] the interior angles on the same side less than two right angles, the two straight lines, if produced indefinitely, meet on that side on which are the angles less than the two right angles.

Euclid states in Definition XXIII (translated by Heath):

Parallel straight lines are straight lines which, being in the same plane and being produced indefinitely in both directions, do not meet one another in either direction.

According to Roberto Bonola, *Non-Euclidean Geometry,* Euclid proves that two straight lines are parallel when they form with one of their transversals either equal interior alternate angles, or equal corresponding angles, or interior angles on the same side that are supplementary, i.e.:

Equal interior alternate angles:

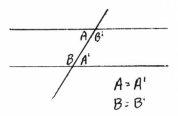

$A = A'$
$B = B'$

Equal corresponding angles:

$A = A'$
$B = B'$

Interior angles on the same side that are supplementary:

$A + B = 180°$
$A' + B' = 180°$

Euclid then puts forth his Fifth Postulate (stated above) and completes his theory of parallels with the following theorems (reworded by Bonola from Heiberg's translation, pp. 1–2):

Straight lines which are parallel to the same straight line are parallel to each other (Bk. I, Prop. 30).

Through a given point one and only one straight line can be drawn which will be parallel to a given straight line (Bk. I, Prop. 31).

The straight lines joining the extremities of two equal and parallel straight lines are equal and parallel (Bk. I, Prop. 33).

For over 2,000 years the task of mathematicians had not been to doubt the certain truth of the postulate but, instead, to show that it could be deduced from Euclid's other postulates. If this deduction were possible, then the Fifth Postulate would be unnecessary, and, if unnecessary, an elimination of it would purify geometry.

In the first century B.C., Posidonius attempted to eliminate the necessity of the postulate by claiming that two equidistant and coplanar straight lines are parallel, i.e., that parallel lines are lines in a single plane which have all perpendiculars equal, whereas lines which have progressively longer or shorter perpendiculars must intersect. Perpendiculars, he says, can determine both the heights of figures and the distances between lines. Therefore, when the lengths of the perpendiculars are equal, the distances between straight lines are equal, but when the lengths of the perpendiculars become longer or shorter, the distance between the lines becomes longer or shorter, and eventually the lines will come together on the side where the perpendiculars become shorter.

However, as Proclus (410–485) pointed out in *A Commentary on the First Book of Euclid's Elements,* first, the absence of intersection does not necessarily make lines parallel, e.g., the circumferences of concentric circles do not intersect; the lines must also extend indefinitely; and second, parallel lines in the Euclidean sense, i.e., lines produced indefinitely which do not meet, exist which would not be parallel under the definition of Posidonius, i.e., equidistant—e.g., a helix inscribed around a straight line, or the asymptotics of a hyperbola or a conchoid.

As Bonola (*Non-Euclidean Geometry,* p. 3) states:

Before we can bring Euclid's definition into line with that of Posidonius, it is necessary to prove that if two coplanar straight lines do not meet, they are equidistant; or, that the locus of points, which are equidistant from a straight line, is a straight line. And for the proof of this proposition Euclid requires his Parallel Postulate.

In the second century, Ptolemy, blinded by the certain validity of the postulate, attempted to deduce it from Euclid's other axioms by asserting that the sum of the internal angles on the same side of a transversal passing through parallel lines is equal to the sum of two right angles. His proof is based on a refutation of the other possibilities, i.e., that the sum of the angles is either greater or less than two right angles. To accomplish this, however, he was forced to *assume* that, if the sum of the internal angles on one side of the transversal is greater or less than two right angles, the same must be true of the sum of the angles on the other side of the transversal.

His proof goes as follows:

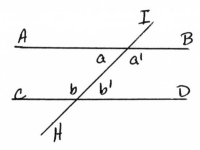

AB and CD are two parallel lines with the transversal HI.

a, b are the two interior angles to the left of HI, and a', b' are the two interior angles to the right of HI.

a + b will be either greater than, equal to, or less than a' + b'.

It is assumed that if this is true for any single pair of parallels it will hold for all others. Thus, if a + b > 2 right angles, it follows that a' + b' > 2 right angles.

Since HB, ID are parallels, and HA, IC are parallels, it follows that a + b > 2 right angles, and thus, that a' + b' > 2 right angles.

Thus, a + b + a' + b' > 4 right angles.

This is *obviously* absurd.

Thus, a + b cannot be greater than 2 right angles.

The same proof works for a + b < 2 right angles.

Therefore, a + b must equal 2 right angles.

Proclus, *assuming* that, by prolonging the lines sufficiently, the distance between two points upon two intersecting straight lines can be made as great as one wishes, deduced that a straight line which meets one of two parallels must meet the other. In order for him to prove this, however, he was forced to assume that the distance between two parallels is *finite*.

Following the work of Proclus until the second half of the sixteenth century, the major attempts to prove Euclid's Fifth Postulate came from Arabian mathematicians, and for the most part they attempted to prove it through some form of equidistance. Aganis, a friend of Simplicius (sixth century), attempted to prove it on the hypothesis that equidistant straight lines exist. Nasir-Eddin (1201–1274) attempted to prove it through the theorem on the sum of the angles of a triangle.

Under the stimulus of the printing of Proclus' *Commentary* in 1533, and a Latin translation of it in 1560, F. Commandino (1509–1575), C. S. Clavio (1537–1612), P. A. Cataldi (?–1626), G. A. Borelli (1608–1679), Giodano Vitale (1633–1711), and others once again took up the attempt to prove Euclid's postulate through the idea of equidistance. They were not successful.

John Wallis (1616–1703) approached the problem from a new direction, the axiom that for every figure there exists a similar figure of arbitrary magnitude.

Walter Prenowitz and Meyer Jordan (*Basic Concepts of Geometry*, pp. 29–31), and Roberto Bonola (*Non-Euclidean Geometry*), offer separate explanations of his proof, which goes as follows:

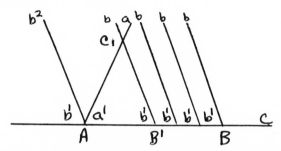

Let a and b be two straight lines intersected at points A and B by the transversal c. Then let angles a' and b' be the interior angles on the same side of c, so that a' + b' = less than 2 right angles.

Draw straight line $b^2$ through A so that b and $b^2$ form equal corresponding angles with c.

Move line b, always keeping the angle equal to b', along line AB, until it intersects with a, forming the triangle AB'C'.

By the hypothesis of similar figures, triangle ABC can now be constructed similar to triangle AB'C'.

Thus, straight lines a and b must meet at a point, the third angle of triangle ABC.

However, the idea of form independent of dimensions is a hypothesis that is no more provable than Euclid's Fifth Postulate and, as Prenowitz and Jordan state, is "hardly less complicated" (p. 31).

Gerolamo Saccheri (1667–1733) made the first significant contribution to an understanding of the Postulate of Parallels. He started with the quadrilateral ABCD where AC≡BD and ∢A and ∢B are right angles:

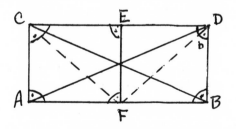

Using congruence, he proved that ∢ACD ≡ ∢BDC. △ABC ≡ △ABD (sas). Thus, AD ≡ BC. Since the diagonals are congruent, △ACD ≡ △BCD (sss). Thus, a ≡ b.

In order to discover whether angles a and b are acute, obtuse, or right angles, it was necessary for Saccheri to determine the sum of the angles of a triangle. He set up the question in the following manner: In triangle ABC (in the following illustration), D and E are the midpoints of AC and BC, and F,G,H are the feet of the perpendiculars C,A,B on the line DE. △DFC and △ADG (saa) are congruent, as are △EFC and △BEH. Thus, the quadrilateral ABHG with right angles at G and H is formed. Therefore, $\angle a + \angle c \equiv \angle b + \angle d$. Since $\angle a + \angle b + (\angle c + \angle d) \equiv$ the sum of the angles in triangle ABC, each of the angles at A and B is equal to the sum of the angles in the triangle divided by two. Thus, the three possibilities are: the hypothesis of the acute angle (the sum of the angles of a triangle is less than two right angles); the hypothesis of the obtuse angle (the sum of the angles of a triangle is greater than two right angles); or the hypothesis of the right angle (the sum of the angles of a triangle is equal to two right angles).

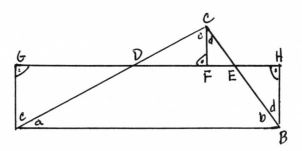

Saccheri was able to eliminate the hypothesis of the obtuse angle by showing that it leads to contradictions. Bonola explains Saccheri's proof of this in the following manner (*Non-Euclidean Geometry,* pp. 34–38):

To make our exposition of Saccheri's work more concise, we take from Prop. XI. and XII. the contents of the following Lemma II:

Let ABC be a triangle of which C is a right angle: let H be the middle point of AB, and K the foot of the perpendicular from H upon AC. Then we shall have

AK = KC, on the Hypothesis of the Right Angle;

AK < KC, on the Hypothesis of the Obtuse Angle;

AK > KC, on the Hypothesis of the Acute Angle.

On the Hypothesis of the Right Angle the result is obvious.

On the Hypothesis of the Obtuse Angle, since the sum of the angles of a quadrilateral is greater than four right angles, it follows that $\angle AHK < \angle HBC$. Let HL be the perpendicular from H to BC:

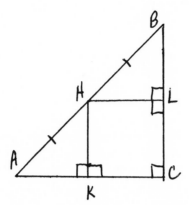

Then the result just obtained, and the fact that the two triangles AHK, HBL have equal hypotenuses, give rise to the following inequality: AK < HL. But the quadrilateral HKCL has three right angles and therefore the angle H is obtuse (Hypothesis of the Obtuse Angle). It follows that HL < KC, and thus AK < KC.

The third part of this Lemma can be proved in the same way.

It is easy to extend this Lemma as follows:

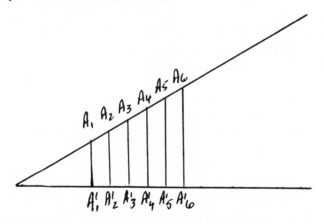

Lemma III. If on the one arm of an angle A equal segments $AA_1$, $A_1A_2$, . . . are taken, and $AA'_1$, $A'_1A'_2$, $A'_2A'_3$, . . . are their projections upon the other arm of the angle, then the following results are true: $AA' = A'_1A'_2 = A'_2A'_3 = $ . . . on the Hypothesis of the Right Angle; $AA'_1 < A'_1A'_2 \ A'_2A'_3 < $ . . . on the Hypothesis of the Obtuse Angle; $AA'_1 > A'_1A'_2 > A'_2A'_3$ . . . on the Hypothesis of the Acute Angle.

To save space the simple demonstration is omitted.

We can now proceed to the proof of Prop. XI. and XII. of Saccheri's work, combining them in the following theorem:

On the Hypothesis of the Right Angle and on the Hypothesis of the Obtuse Angle, a line perpendicular to a given straight line and a line cutting it at an acute angle intersect each other:

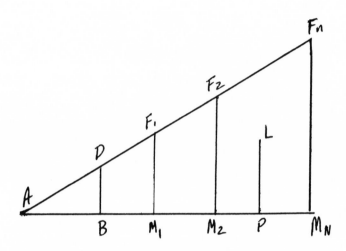

Let LP and AD be two straight lines of which the one is perpendicular to AP, and the other is inclined to AP at an acute angle DAP.

After cutting off in succession equal segments AD, $DF_1$, upon AD, draw the perpendiculars DB and $F_1M_1$ upon the line AP.

From Lemma III. above, we have $BM_1 \geqq AB$, or $AM_1 \geqq 2AB$, on the two hypotheses.

Now cut off $F_1F_2$ equal to $AF_1$, from $AF_1$ produced, and let $M_2$ be the foot of the perpendicular from $F_2$ upon AP. Then we have $AM_2 \geqq 2AM_1$, and thus $AM_2 \geqq 2^2AB$. This process can be repeated as often as we please.

In this way we would obtain a point $F_n$ upon the line AD such that its projection upon the line AP would determine a segment $AM_n$ satisfying the relation $AM^n \geqq 2^nAB$.

But if n is taken sufficiently great, (by the Postulate of Archimedes) we would have $2^nAB > AP$, and therefore $AM_n > AP$. Therefore the point P lies upon the side $AM_n$ of the right angled triangle $AM_nF_n$. The perpendicular PL cannot intersect the other side of this triangle; therefore it cuts the hypotenuse.

It is now possible to prove the following theorem:

The Fifth Postulate is true on the Hypothesis of the Right Angle and on the Hypothesis of the Obtuse Angle (Prop. XIII).

Let AB, CD be two straight lines cut by the line AC:

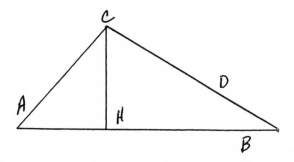

Let us suppose that ∡BAC + ∡ACD < 2 right angles.

Then one of the angles ∡BAC, ∡ACD, for example the first, will be acute.

From C draw the perpendicular CH upon AB. In the triangle ACH, from the hypotheses which have been made, we shall have ∡A ≠ ∡C ≠ ∡H ≧ 2 right angles. But we have assumed that ∡BAC ≠ ∡ACD < 2 right angles. These two results show that ∡AHC > ∡ HCD. Thus the angle HCD must be acute, as H is a right angle. It follows from Prop. XI., XLL. that the lines AB and CD intersect.

This result allows Saccheri to conclude that the Hypothesis of the Obtuse Angle is false (Prop. XIV.).

Thus, Saccheri was able to show that the sum of the angles of a triangle does not exceed two right angles.

However, the most important aspect of all of Saccheri's labors is that they *failed* to show that the hypothesis of the acute angle is false. Though it lay directly before his eyes, Saccheri, blinded by an assumption (that the Fifth Postulate was correct), could not see what he had done.

Johann Lambert (1728–1777) carried Saccheri's studies to a point where preconceived ideas become amazing. He, going so far as to suggest that this may be the same geometry which would occur upon a sphere, actually began to discover the laws of a geometry based on the hypothesis of the acute angle, e.g., the defect of a polygon and the absolute unit for geometric segments.

Adrien Legendre (1752–1833) gave a very elegant proof of Saccheri's rejection of the hypothesis of the obtuse angle. He then built upon this and carried it to what he thought was a proof of the Postulate of Parallels. However, once again there is a problem. In order for his proof to work, it is necessary to *assume* that it is always possible to draw a line through a point in the interior of an angle which intersects both sides of the angle.

By 1763, G. S. Kluegel was able to find and reject twenty-eight significant attempts to prove the postulate. So elusive was its proof that mathematicians had finally begun to accept it as an axiom. Nevertheless, its intuitive appeal was so strong that an acceptance of its unprovableness did not lead to a questioning of its validity.

Wolfgang Bolyai (1775–1856), once again blinded by an absolute obedience to the truth of the postulate, continued the search for its proof. Although unable to prove it, he came up with an important new approach to the problem, e.g., four points, not in a plane, always lie upon a sphere; or, a circle can always be drawn through three points not on a straight line. He, however, became so frustrated with the task that he gave up on it, and, deciding that it was something which one has to accept on faith, strongly advised his son, Johann Bolyai, to stay clear of it.

Karl Friedrich Gauss (1777–1855), although he did not publish his proofs, was the first to grasp the idea of a geometry *independent of* Euclid's Fifth Postulate. Not only did he grasp the idea, but within his notes the problems of a non-Euclidean geometry are solved. The most important of these is the solution of an absolute unit of length:

$$\pi k \left( \frac{r}{ek} - e\frac{r}{k} \right), \quad k = \infty$$

Yet either he was too afraid of its implications, or was unsure of his calculations, or did not realize the magnitude of what he had done to pursue it.

Ferdinand Karl Schweikart (1780–1859) also discovered a geometry (astral geometry) independent of Euclid's Fifth Postulate. He communicated his findings to Gauss, and the following formula was developed:

$$\frac{\pi cc}{[\log \, hyp(1+\sqrt{2})]^2}, \quad k = \frac{c}{\log(1+\sqrt{2})}$$

Once again, the findings were not published.

Franz Adolf Taurinus (1794–1874) developed logarithmic-spherical geometry, which is based on the Hypothesis of the Acute Angle. Yet he could not take it seriously: it was a world of make-believe.

In 1829 Nikolas Ivanovich Lobatchewsky (or Lobachevski; 1793–1856) published Plangeometry, a geometry which, containing no contradictions, is as logical as Euclid's. Building on the formula $\frac{\cos A \, \cos\pi(b) \, \cos\pi(c) \, + \, \sin\pi(b)\sin\pi(c)}{\sin\pi(a)} = 1$, where the sides are denoted by a,b,c, the angles by A,B,C, and the angles of parallelism corresponding to the sides a,b,c by $\pi(a)$, $\pi(b)$, $\pi(c)$ (this is easily translated into Taurinus' formula), he was able to create a trigonometry which covers the Hypothesis of the Right Angle (Euclid's geometry), the Hypothesis of the Obtuse Angle, and the Hypothesis of the Acute Angle. This trigonometry enabled him to create a hyperbolic geometry where the sum of the angles of a triangle is always less than two right angles, there are two parallels (asymptotic lines) from an exterior point to a given line, a horocycle (circle of infinite radius) takes the place of Euclid's straight line, and a horosphere (sphere of infinite radius) takes the place of Euclid's plane.

In 1833, even though his father had strongly discouraged him, Johann Bolyai (1802–1860), independent of Lobatchewsky, also put forth a non-Euclidean geometry. His is based on the following formula, which is the key to all non-Euclidean trigonometry:

$$e - \frac{g}{k} = \tan \frac{\pi (a)}{2}$$

By working out his formulae for absolute trigonometry, it is once again possible to demonstrate a system of mathematics which works for the Hypothesis of the Right Angle, the Hypothesis of the Obtuse Angle, and the Hypothesis of the Acute Angle.

What Schweikart, Taurinus, Gauss, Lobatchewsky, and Bolyai had proven, by constructing other geometries which were as logical as Euclid's, was that Euclid's Fifth Postulate is an axiom, i.e., a statement accepted without proof on no other grounds than that it seems to be correct. They did this by reversing the question. If the postulate could be reduced, i.e., is not an axiom, a reversal of the postulate would create contradictions within Euclidean geometry. Retaining all of the other axioms of Euclidean geometry, Lobatchewsky changed the Fifth Postulate to state that through a given point two parallels can be drawn to a given line. Accepting this as the new postulate, he was able to construct a new geometry which, containing no contradictions, is as logical as Euclid's. Thus, the Fifth Postulate cannot be reduced to simpler terms: it is an axiom. Therefore, Euclid's geometry, considered the only *correct* geometry for centuries, is neither true nor false but only the logical development of an *arbitrary* set of axioms.

It would be hard to overestimate the magnitude of what Lobatchewsky (he is the major figure) did. E. T. Bell, *Men of Mathematics,* expresses it in the following manner:

Instead of constructing a geometry to fit the Earth as human beings now know it, Euclid apparently proceeded on the assumption that the Earth is flat. If Euclid did not, his predecessors did, and by the time the theory of "space," or geometry, reached him the bald *assumptions* which he embodied in his postulates had already taken on the aspect of hoary and immutable necessary truths, revealed to mankind by a higher intelligence as the veritable essence of all material things. It took over two thousand years to knock the eternal truth out of geometry, and Lobatchewsky did it.

To use Einstein's phrase, Lobatchewsky *challenged an axiom.* Anyone who challenges an "accepted truth" that has seemed necessary or reasonable to the great majority of sane men for 2000 years or more takes his scientific reputation, if not his life, in his hands. Einstein himself challenged the axiom that two events can happen in *different places* at the *same time,* and by analyzing this hoary assumption was led to the invention of the special theory of relativity. Lobatchewsky challenged the assumption that Euclid's parallel postulate or, what is equivalent, the hypothesis of the right angle, is necessary to a consistent geometry, and he backed his challenge by producing a system of geometry based on the hypothesis of the acute angle in which there is not *one* parallel through a fixed point to a given straight line but *two.* Neither of Lobatchewsky's parallels meets the line to which both are parallel, nor does any straight line drawn through the fixed point and lying within the angle formed by the two parallels. This apparently bizarre situation is "realized" by the geodesics on a pseudosphere.

For any everyday purpose (measurements of distances, etc.), the differences between the geometries of Euclid and Lobatchewsky are too small to count, but this is not the point of importance: each is self-consistent and each is adequate for human experience. Lobatchewsky abolished the *necessary* truth of Euclidean geometry. His geometry was but the first of several constructed by his successors. Some of these substitutes for Euclid's geometry—for instance the Riemannian geometry of general relativity—are today at least as important in the still living and growing parts of physical science as Euclid's was, and is, in the comparatively static and classical parts. For some purposes Euclid's geometry is best or at least sufficient, for others it is inadequate and a non-Euclidean geometry is demanded.

Euclid in some sense was believed for 2200 years to have discovered an absolute truth or a necessary mode of human perception in his system of geometry. Lobatchewsky's creation was a pragmatic demonstration of the error of this belief. The boldness of his challenge and its successful outcome have inspired mathematicians and scientists in general to challenge other "axioms" or accepted "truths," for example the "law" of causality, which, for centuries, have seemed as necessary to straight thinking as Euclid's postulate appeared till Lobatchewsky discarded it. (pp. 305–306)

Of even more importance is the influence of Lobatchewsky's discovery beyond the disciplines of mathematics and science. Immanuel Kant, the most important philosopher at the end of the eighteenth century, in a sense based his entire philosophy on the absolute validity of Euclidean geometry. Morris Kline (*Mathematical Thought from Ancient to Modern Times,* p. 862) states:

Kant's answer to the question of how we can be sure that Euclidean geometry applies to the physical world, which he gave in his *Critique of Pure Reason* (1781), is a peculiar one. He maintained that our minds supply certain modes of organization—he called them intuitions—of space and time and that experience is absorbed and organized by our minds in accordance with these modes or intuitions. Our minds are so constructed that they compel us to view the external world in only one way. As a consequence certain principles about space are prior to experience; these principles and their logical consequences, which Kant called a priori synthetic truths, are those of Euclidean geometry. The nature of the external world is known to us only in the manner in which our minds oblige us to interpret it. On the grounds just described Kant affirmed, and his contemporaries accepted, that the physical world must be Euclidean. In any case whether one appealed to experience, relied upon innate truths, or accepted Kant's view, the common conclusion was the uniqueness and necessity of Euclidean geometry.

Thus, by showing that Euclidean geometry is built upon a subjective foundation, it was possible to bring the entire structure of human knowledge into question.

Contemporary mathematicians, then, have three courses of action in reference to Euclid's Fifth Postulate. They can continue to attempt to prove it, even though it has been proven that it cannot be proved. This sounds ridiculous, but then some mathematicians are still tempted to trisect an angle with a compass and straight-edge (though that has also been proven impossible), and after all, aren't advances in mathematics always the result of challenging an absolute?

They can join in on the development of non-Euclidean geometries, as did Bernhard Riemann, Hermann Ludwig von Helmholtz, and Eugenio Beltrami, and the projective geometries of such men as Felix Klein and Arthur Cayles.

Or they can explore the philosophical implications, in which case David Hilbert's *Grundlagen der Geometrie* (1899; trans. as *The Foundations of Geometry,* 1902), removing geometry from sensory space, becomes a central document.

This clear cut between mathematics and sensory experience becomes a central aspect of twentieth-century mathematics, as expressed in the following lines from "Geometrie und Erfahrung," a lecture given by Albert Einstein in 1921 (taken from "The History of Mathematics," in *Encyclopaedia Britannica,* 1985):

As far as the mathematical theorems refer to reality, they are not sure, and as far as they are sure, they do not refer to reality. . . . The progress entailed by axiomatics consists in the clean-cut separation of the logical form and the realistic and intuitive contents. . . . The axioms are voluntary creations of. . . human mind. . . . To this interpretation of geometry I attach great importance for should I not have been acquainted with it, I would never have been able to develop the theory of relativity.

For similar explorations and recreations refer to The Duplication of the Cube; Geometric Problems and Puzzles; Squaring the Circle; The Trisection of an Angle.

BIBLIOGRAPHY

Armstrong, James W. *Elements of Mathematics*. New York: Macmillan and Co., 1970.

Ball, W. W. Rouse. *A Short Account of the History of Mathematics*. 1888. Reprint. London: Macmillan and Co., 1935.

Banks, J. Houston. *Elements of Mathematics*. Boston: Allyn and Bacon, 1961.

Behnke, H., F. Bachmann, K. Fladt, and H. Kunle. *Fundamentals of Mathematics. Vol. II: Geometry*. Trans. by S. H. Gould. Cambridge, Mass.: MIT Press, 1974.

Bell, E. T. *Men of Mathematics*. New York: Simon & Schuster, 1937.

Blumenthal, Leonard M. *A Modern View of Geometry*. San Francisco: W. H. Freeman and Co., 1961.

Bonola, Roberto. *Non-Euclidean Geometry*. Trans. by H. S. Carslaw. Chicago: Open Court Publishing Co., 1912.

Born, Max. *Experiment and Theory in Physics*. Cambridge, Mass., 1943. Reprint. New York: Dover, 1956.

———. *Physics in My Generation*. New York: Springer-Verlag, 1969.

Calinger, Ronald, ed. *Classics of Mathematics*. Oak Park, Ill.: Moore Publishing Co., 1982.

D'Abro, A. *The Evolution of Scientific Thought from Newton to Einstein*. New York: Boni and Liveright, 1927.

Dubbey, J. M. *Development of Modern Mathematics*. New York: Crane, Russak & Co., 1970.

Einstein, Albert. "Geometrie und Erfahrung." 1921. Reprinted in "The History of Mathematics." *Encyclopaedia Britannica,* 1985.

Eves, Howard. *An Introduction to the History of Mathematics*. 1953. Reprint. New York: Holt, Rinehart & Winston, 1969.

Heiberg, J. L. *The Thirteen Books of Euclid's Elements*. Leipzig: Teubner, 1883. Trans. by Sir Thomas L. Heath. Cambridge, Eng.: Cambridge University Press, 1908.

Heisenberg, Werner. *Physical Principles of the Quantum Theory*. Trans. by C. Eckart and F. C. Hoyt. Chicago: University of Chicago Press, 1930. Reprint. New York: Dover, 1949.

Keysler, Cassius Jackson. *The Pastures of Wonder*. New York: Columbia University Press, 1929.

Kline, Morris. *Mathematical Thought from Ancient to Modern Times*. New York: Oxford University Press, 1972.

LeVeque, William Judson, et al. "The History of Mathematics." In *Encyclopaedia Britannica,* 1985.

Manning, Henry Parker. *Non-Euclidean Geometry*. 1901. Reprint. New York: Dover, 1963.

Meschkowski, Herbert. *Noneuclidean Geometry*. Trans. by A. Shenitzer. New York: Academic Press, 1964.

Meserve, Bruce E. *Fundamental Concepts of Geometry*. Reading, Mass.: Addison-Wesley Publishing Co., 1955.

Poincaré, Henri. *The Foundations of Science*. Trans. by George Bruce Halsted. New York: Science Press, 1913.

_____. *Mathematics and Science: Last Essays.* 1913. Reprint. Trans. by John W. Bolduc. New York: Dover, 1963.

Prenowitz, Walter, and Meyer Jordan. *Basic Concepts of Geometry.* Lexington, Mass.: Xerox, 1965.

Proclus. *A Commentary on the First Book of Euclid's Elements.* Trans. by Glenn R. Morrow. Princeton, N.J.: Princeton University Press, 1970.

Russell, Bertrand A. W. *An Essay on the Foundations of Geometry.* 1897. Reprint. New York: Dover, 1956.

Schaaf, William, ed. *Our Mathematical Heritage.* New York: Collier, 1963.

Shaw, James Byrnie. *The Philosophy of Mathematics.* Chicago: Open Court, 1918.

**EULERIAN SQUARES.** See MAGIC SQUARES.

# F

**FAIRY CHESS** is a general term for variations on chess (*see* Chess) involving changes in the rules of play, the form of the board, or the pieces used. It is related to and may serve as the basis for chess problems (*see* Chess Problems), but remains a game to be played by two players. Chess problems, on the other hand, though they may involve the rules of fairy chess, in which case they might more accurately be called fairy chess problems, are mathematical puzzles to be solved.

Legend has it that chess originated in India about A.D. 600. However, it did not solidify into the standard modern version until about 1500. Even then, numerous variations existed throughout Europe concerning castling, pawn promotion, stalemate, *en passant* captures, and so on. It was not until as recently as 1924 that the Fédération Internationale des Echecs (FIDE) established an official set of rules for international tournaments. Since then, all of the variations from this official set of rules fall under the designation of fairy chess.

A great outburst of activity in fairy chess and chess problems began about 1918, led by the great problem composer T. R. Dawson (1889–1951), and resulted in the establishment of a Heterodox (fairy chess) Section in the FIDE Problem Section, and in inclusion in the Codex drawn up after the International Congress of Problemists at Piran, Yugoslavia, under the auspices of FIDE in 1958.

This section recognizes and establishes as standard fairy chess pieces the grasshopper and the nightrider. It also establishes two fairy chess rule variations, selfmate (where White must force Black to mate White in *n* number of moves) and helpmate (where Black moves to help White mate Black in *n* moves). It extends fairy chess recognition to certain chessboards that vary from the standard chessboard, notably the vertical cylinder. And it includes as a standard fairy chess game the Maximummer, a game invented by Dawson where Black must always play its geometrically longest move.

However, extensive as recent activities in fairy chess have been, variations on standard chess—what would now be called fairy chess variations—were taking place long before 1918. H. J. R. Murray, *A History of Chess,* describes a number of these earlier forms, including oblong chess, decimal chess, astronomical chess, and limb chess.

Great chess or complete chess, played on an 11 × 10 square board, is supposed to have been a favorite game of the Mongol Emperor, Timur (1336–1405). This form added a number of pieces to the standard chessmen: the wazir, the dabbaba, the talia, the camel, and the giraffe.

Other forms derived from great chess included such pieces as the lion, the bull, the sentinel, the crocodile, the paymaster, the prince, the chief of police, the rhinoceros, and the gazelle.

The courier game, played in Europe in the twelfth and thirteenth centuries on a 12 × 8 board, included two couriers on each side, a sneak on each side, and a man on each side, in addition to the standard pieces. (*See* Chess for a more in-depth discussion of the history, including other variations.)

The following are some of the more common fairy chess pieces:

The *grasshopper* moves in the same way as the queen, except that it must hop over a piece (of either color) and land on the square beyond.

The *locust* is derived from the grasshopper, but it must hop over a man of the opposite color to the square beyond, which must be empty, capturing the man it hops over.

The *lion* is also derived from the grasshopper, but it must hop over a man of either color to move or capture, landing on any square in the line beyond that man.

The *fers* moves one square diagonally. If it results from the promotion of a pawn, it may also jump one square in any direction. It is sometimes referred to as a one-step leaper.

The *wazir* moves one square orthogonally. It too is sometimes referred to as a one-step leaper.

The *alfil* moves diagonally, like a standard bishop (it is an earlier form of the bishop), but it can only move two squares at a time. It does not matter if the intervening square is occupied. It is sometimes referred to as a two-step leaper.

The *dabbaba* moves on the orthogonals in the same manner that the alfil moves on the diagonals, and is also referred to as a two-step leaper.

The *camel* moves either three spaces vertically and one laterally or three spaces laterally and vertically. It does not matter if the intervening squares are occupied. It is sometimes referred to as a three-step leaper.

The *nightrider* moves with multiple knight moves in the same pattern (same line) as its first move until it lands on a square occupied by another piece or is stopped by the end of the board. The knight is a standard chess piece, a two-step leaper that moves two spaces vertically and one laterally or two spaces laterally and one vertically.

The *camelrider* moves in the same manner as a nightrider, but on a line built up of three-step moves.

The *dabbaba-rider* moves in multiple leaps along an orthogonal line two spaces at a time. It leaps over intervening pieces if necessary.

The *alfil-rider* moves in the same manner as a dabbaba-rider, except that it moves diagonally.

The *dabbaba-riderhopper* moves in the same manner as a dabbaba-rider, except that the step directly before it lands must be occupied by a chess piece of either color which it must hop over.

The *nightrider-hopper* moves in the same manner as a nightrider, except that, as with the dabbaba-riderhopper, it must hop over a square occupied by a chess piece of either color which rests on the square the nightrider-hopper would normally occupy if it stopped one step earlier.

The *camel-riderhopper* combines the moves of the camelrider and the hopper in the same manner as the dabbaba-riderhopper and the nightrider-hopper.

The *alfil-riderhopper* combines the moves of the alfil-rider and the hopper in the same manner as the dabbaba-riderhopper and the nightrider-hopper and camel-riderhopper.

The *rookhopper* combines the moves of the rook, a standard chess piece that moves along orthogonal lines, with the moves of a hopper in the same manner that the grasshopper adds a hop to the standard moves of a queen.

The *bishopper* combines the moves of the bishop, a standard chess piece that moves along diagonal lines, with the moves of a hopper in the same manner that the grasshopper adds a hop to the standard moves of a queen.

The *giraffe* moves either four spaces vertically and one laterally or four spaces laterally and one vertically. It is sometimes referred to as a four-step leaper.

The *zebra* moves either three spaces vertically and two laterally or three spaces laterally and two vertically.

The *fiveleaper* moves in accordance with a distance of $\sqrt{25}$, which translates into a move of five spaces along either orthogonal line, or four spaces vertically and three laterally, or four laterally and three vertically.

The *root-50-leaper* moves in accordance with a distance of $\sqrt{50}$, which translates into a move of seven spaces vertically and one laterally, a move of seven spaces laterally and one vertically, or a move of five vertically and five horizontally.

The *girafferider* moves in the same manner as the giraffe, except that it gains the same multiple move capacity of other riders. It is only possible on a board of larger than standard dimensions.

The *zebrarider* moves in the same manner as the zebra, except that it gains the same multiple move capacity of other riders.

The *rank-rider* moves in the same manner as a rook or castle, except that it can only move laterally.

The *filerider* moves in the same manner as a rook or castle, except that it can only move vertically.

The *demibishop* or *half-bishop* moves in the same manner as a bishop, except that it can only move on one diagonal.

The *grandbishop* or *great bishop* moves along prolongations of the diagonals (important on a vertical cylinder).

The *archbishop* moves in the same manner as a bishop with the added ability to bounce off one edge of the board in the manner of a billiard ball.

The *reflecting bishop* moves in the same manner as a bishop, but may reflect off all edges of the board.

The *edgehog* moves like a queen, but only from or to a square on the edge of the board.

The *leo* moves like a queen, except when it captures a piece, in which case it must hop over that piece to any square beyond on the same line.

The *pao* moves like a rook, except when it captures a piece, in which case it must hop over that piece to any square beyond on the same line.

The *vao* moves like a bishop, except when it captures a piece, in which case it must hop over that piece to any square beyond on the same line.

The *nighthopper* moves like a nightrider-hopper, but must stop on the second leap of its riding-line.

The *rankhopper* is a rook-or-castlehopper confined to lateral movement.

The *filehopper* is a rook-or-castlehopper confined to vertical movement.

The *equihopper* combines the moves of all the hoppers. However, it must hop over a piece of either color that rests at the exact center of its move. Because of this, the equihopper always remains on the same color square it begins the game on.

All of the above pieces are either leapers, hoppers, or riders, or a combination of two. Leapers are pieces which leap, as does the knight, to various lengths of the *a-b* formula (the knight's move would be either 1,2 or 2,1). Riders are leapers that have the power of extending their first leap in the same direction by taking additional leaps along a "riding-line." Hoppers are riders which *must* hop over a piece in the process of making a move.

The following are some of the fairy pawns:

The *Berolina pawn* moves diagonally, but captures orthogonally. It may capture another Berolina pawn *en passant* (this occurs when a pawn, in moving two squares, passes the diagonal of an opponent's pawn, in which event the latter may take it in passing by moving on to the square it crossed).

The *dummy* or *moveless pawn* may neither move nor capture, but it may be captured.

The *latent pawn* has crossed the board and awaits a piece to be captured to be promoted to that piece's rank.

The *neutral pawn* may be used by either player. It is regarded as whatever color the player whose turn it is wishes it to be.

The *protean pawn* may (but does not have to) assume the rank of any piece it captures.

The *retreatable pawn* plays between the second and eighth ranks of the board for White or between the seventh and first ranks of the board for Black. It moves forward as a regular pawn but also moves in reverse. It can only capture on a standard forward move, but it may be captured *en passant* on its backward double move. It may either promote naturally or remain a retreatable pawn.

The *reversible pawn* moves, captures, and checks backward as well as forward.

The *superpawn* may extend its normal move to any distance. It captures at any point along its diagonals.

The following pieces have not been derived from normal chessmen and might be called supernumerary pieces:

The *imitator* cannot check, capture, or be captured. It moves in exact imitation of the moves made by each piece of either color. It cannot enter an occupied square. Therefore, if a piece is moved that would force the imitator to enter an occupied square, then the move is illegal.

The *joker* takes on the powers of the last man moved by the opponent. It may be either a Black or a White piece and has full moving, checking, and capturing powers.

The *diplomat* does not move. It cannot capture or be captured. It saves from capture every friendly man adjacent to it.

The *pyramid* remains immovable on a square, which it blocks for all other men of either color. It cannot be captured.

The following pieces combine the moves of two or more other pieces:

The *omnipotent queen, terror, general,* or *amazon* combines the moves of the queen and the knight.

The *chancellor* or *empress* combines the moves of the rook and the knight.

The *centaur* or *princess* combines the moves of the bishop and the knight.

The *gnu* combines the moves of the camel and the knight.

The *dragon* combines the moves of the pawn and the knight.

The *gryphon* or *griffin* combines the moves of the pawn and the bishop.

The *squirrel* combines the moves of the alfil and the dabbaba and the knight.

The *demi-queen* combines the moves of the rook and the demi-bishop.

The *grand queen* combines the moves of the rook and the grand-bishop.

It is possible to combine the powers of movement of one piece with the powers of capture of another piece, e.g., bishop (mover)/rook (capturer). It is also possible for a piece to move forward in the manner of one type of chess piece and backward in the manner of a different type. Such pieces are generally called *hunters.* An *R/B hunter,* for instance, would move forward as a rook and backward as a bishop.

The following pieces are less commonly used:

The *arrow bishop* has all the normal powers of a bishop with the additional power, when checking, of also controlling the two adjacent orthogonal squares at its arrowhead. For example, if the bishop at a-1 (the lower left corner square of the chessboard) checks the king at e-5, it also controls e-4, d-4, d-5, and f-6.

The *arrow pawn* may move one or two steps in any orthogonal direction and capture on any diagonal. It may not be promoted. It may be captured *en passant* on any double step.

The *arrow rook* moves like a rook, but if it is on the same rank or file as the opponent's king, then it also controls the two diagonal squares the king might otherwise move into of the rank or file toward the rook.

The *atomic bomb* can be attained by a pawn promotion and then dropped onto any square, where it destroys everything around that square for an area of $n$ steps. If it destroys the king, then the next highest ranking piece becomes king. It can only be played once.

The *balloon* is a bishop in four-dimensional chess.

The *boyscout* is a rider that moves on a zigzag diagonal.

The *cadet* is a pawn that has crossed the board before any higher ranking pieces of its color have been captured. It waits for a piece to be captured so that it can be promoted.

The *capturing pieces* may only move in order to capture. The *Atlantosaurus* may only move to capture the king. The *dinosaurus* may only move to capture the queen. The *mammoth* may only move to capture a rook. The *brontosaurus* may only move to capture a bishop. The *hippopotamus* may only move to capture a knight. The *polyp* may only capture a piece on the queen's lines (it has no effect on the Black king).

The *chameleon* changes its powers on each successive move by a player, starting as a knight, then a bishop, then a rook, then a queen, and then back to a knight.

The *Circe piece,* when captured, is placed back on its square of origin—in the case of a rook, a bishop, or a knight, on the square of the same color as it was captured on; in the case of a pawn, on the initial square of its file. If the square of origin is occupied, the piece is removed from the board.

The *cowboy* is a noncapturing rider on the perimeter squares of the board. It "lassos" any man on its rank or file (in the center of the board) that has eight vacant squares around it, stalemating that piece.

The *cyclic pieces* are used on a standard chessboard, but have the power pieces would normally have on a vertical cylinder. They cannot, however, return to their original starting square.

The *double-grasshopper* must make two consecutive grasshopper hops; change or reversal of direction is permitted.

The *double-knight* makes two consecutive knight moves (it can stop on the first). Its second leap must be away from its starting position, but not to a square that a nightrider could visit.

The *double-move queen* makes a bishop move followed by a rook move, or a rook move followed by a bishop move.

The *equigrasshopper* is an equihopper limited to ranks, files, and diagonals.

The *griffon* (not the *griffin*) moves like a knight, except when it lands on a square of its own color, in which case it may make an additional knight move.

The *halma-grasshopper* makes *n* consecutive grasshopper moves with change or reversal of direction permitted through empty landing squares.

The *hydra* is gained only if a pawn reaches promotion rank before any pieces of its own color have been captured. It is a knight with a double-leap.

The *invisible man* may be revealed, assuming the powers of any fairy piece currently on the board or of the queen, rook, bishop, knight, or pawn, on any vacant square on the board from which it does not place a king in check. It must then make a move (which may result in a check). If a player has no other move, he must reveal his invisible man. If it is captured, it is permanently killed; if it is not captured immediately after its first move, it becomes invisible at once and may be used again. If a pawn promotes into an invisible man, it immediately becomes invisible. While invisible, the invisible man at the same time may be either intangible (pieces may move through its square or even land on its square) or tangible (in which case other pieces may not move across or occupy its square).

The *jibber* moves like a queen but stops short of the first man it meets.

The *joker king,* after each move, takes on the move a guard power of the opposite player's man moved.

The *kamikaze piece* disappears from the board along with the first piece it captures.

The *kangaroo* moves like a grasshopper, except that it moves to the next square beyond two men which need not stand adjacent to each other.

The *laser piece* must stop short of the square occupied by an enemy piece, at which time it emits a laser-beam that destroys pieces of either color (the first piece along each of its lines of control).

The *minotaurus* moves only on the 2-square diagonals in the corners of the board (e.g., a2-b1).

The *mermaid* moves like a queen, but to capture it must hop over an opponent's man to the next square beyond to capture that man.

The *neutral pieces* may be played by either player. The following rules apply: Neither player may make a move leaving his king in check from a neutral piece. Any neutral piece may be moved or captured by either player, including the capture of a neutral piece by a neutral piece. White moves neutral pawns up the board; Black moves them down. Mate by a neutral piece is only possible when the mating move is in some way irreversible.

The *nostalgic king* is a king that, upon reaching a square a queen's move away from its home square, must move home on the next move.

The *overrunner* moves like a bishop, except that upon reaching the edge of the board it may move one orthogonal square along the edge at an angle of 45 degrees.

The *protean king* acquires the rank of each piece captured, until the next capture alters it.

The *Red-Cross men* cannot capture, but are also immune from capture. They move only to interpose by the quickest route on a line of check.

The *royalties* are kings for checks, checkmates, and stalemates, but move under their own powers for all other moves.

The *serpent* moves to all nightrider squares and their single diagonal outgoing squares.

The *tank* moves like a king, except that it pushes its own colored pieces onward to the next square. As many pieces as are lined up in a straight line may be pushed onward. If an opponent's piece is in the line, it is captured. Pawns may be pushed back to their starting position and regain the option of a double move.

The *transparent pieces* allow other pieces to pass through them.

The *trizebra* is a sixth-step leaper. It makes three consecutive zebra leaps forming an "N" shape.

The *unicorn* is a three-dimensional bishop in space chess.

The *vaulting kings* are kings which take on the power of some other piece when in check.

The *victory hoppers* move like equihoppers on a line of even leaps, bent halfway to form the shape of a "V".

The *victory leapers* form a "V" pattern on their second leap.

The *x-ray pieces* may move through any piece.

The *zeppelin* is a rider that moves along every curve whose successive steps are: $(1,x)$ $(1,x+1)$ $(1,x+2)$ . . .

The *zero* may only jump straight up and land back on the same square.

Numerous variations of the standard chessboard have been developed. Former chess champion José R. Capablance introduced a board of $12 \times 16$ squares, using 32 pieces on each side. H. D. Baskerville created a game to be played on a rectangular arrangement of 83 hexagonal spaces. W. Stead invented grid-chess, where the chessboard is divided into 16 squares of 4 units each, and every move by any man must cross at least one of the grid lines.

Perhaps the most interesting of the chessboard variations is the cylindrical board. There are three principal types: the vertical cylinder, the horizontal cylinder, and the anchor-ring. The vertical cylinder is a board (or, rather, an imagined board) that wraps around a cylinder, so that the left and right edges come together. Thus, a piece is able to move off one edge of the board and reenter the board on the other edge. The horizontal board wraps around a cylinder vertically. There are no upper or lower edges. The anchor-ring combines the vertical cylinder and the horizontal cylinder. Unless otherwise stated, every move on a cylinder board must be finite.

Chessboards of three or more dimensions result in a form of chess called space chess. The most common form takes place on a 5 × 5 × 5 space cube of 125 cells. The pieces for the standard form of this are the normal chess pieces, plus the unicorn (described above). Each man occupies a cell. The rook has the standard rook moves, plus vertical orthogonal moves. The bishop moves through cell edges (e.g., Aa1 to Ab2). The unicorn moves through cell corners (e.g., Aa1 to Bb2). The queen combines the moves of the rook, the bishop, and the unicorn. The king moves in single-step moves in the same manner as the queen. The knight moves like a standard knight, plus vertical knight moves. Pawns move rook-wise in single steps only and capture bishop-wise, always toward their promotion rank. The initial set-up is as follows. For White: king—Ac1; queen—Bc1; rooks—Aa1 and Ae1; Bishops—Ba1 and Bd1; unicorns—Bb1 and Be1; knights—Ab1 and Ad1; pawns—Aa2, Ab2, Ac2, Ad2, Ae2, Ba2, Bb2, Bc2, Bd2, Be2; and the pawns promote on E5 rank. For Black: king—Ec5; queen—Dc5; rooks—Ea5 and Ee5; bishops—Db5 and De5; unicorns—Da5 and Dd5; knights—Eb5 and Ed5; pawns—Ea4, Eb4, Ec4, Ed4, Ee4, Da4, Db4, Dc4, Dd4, De4; and the pawns promote on A1 rank.

Four-dimensional boards allow each piece one more possible direction. The boards can be represented by a diagram of four boards horizontally and four boards vertically. One extra piece is usually added (the balloon, explained earlier).

Oblique cylinder boards are imagined to be wrapped obliquely around a cylinder, so that the diagonals join, instead of ranks or files. The spherical board is imagined to be wrapped around a sphere, its squares formed by lines of latitude and longitude, as if it were a globe. Möbius bands (*see* The Möbius Strip) may serve as a chessboard, as may Möbius rings (an anchor-ring with a 180 degree twist, so that first-rank squares a1-d1 join eighth-rank squares e8-h8).

Generally, when the rules of play are changed from those of standard chess, it is to give a handicap to the stronger player. The changes may involve giving the weaker player the advantage of a move, or an extra piece and a move, or some such combination. Sometimes certain moves are made obligatory. In the Marseille Game, each player plays two single moves in succession. Check may be given only with the second move, and the first move of the player in check must be to move out of check.

The Legal Game requires the player with the White queen to use eight extra pawns instead of the queen. The Marked-Pawn Game is a strongly handicapped game where a pawn (usually the king's knight's pawn) is marked. If this pawn is captured, its owner loses the game. This player may also lose the game in the standard way by having his king checkmated, but he may only win the game by checkmating his opponent with the marked pawn. The marked pawn may not be moved across the board and exchanged for another piece.

Kriegspiel requires each player to have a separate board with only his pieces on it. The umpire has a third board and tells the players whether their suggested moves are legal or not. The players are not allowed to see each other's board. A player may ask on his move "Are there any?" meaning are there any pawn captures. If the umpire says "Yes," then the player must attempt a pawn capture. If his move is successful, then he makes the capture and his move is over. If the capture is not playable, then he continues with other moves until a legal move is made. The umpire also announces check on the rank, file, long diagonal, or short diagonal—nothing more. If the check is made from a knight, the umpire simply says "Check from Knight." When the umpire announces a capture, he only says which square the capture is made on (announcing neither the piece captured nor the piece making the capture).

Checkless chess requires that neither side may place the other in check, except to give mate. Black Must Check requires Black to give check if he can. Neither-Side-May-Capture is just what the name says it is. Black-Moves-Only-To-Capture requires Black not to move at all, unless he can capture a piece, in which case he must do so. White may only check if Black has a capturing move in reply. Black-Moves-Only-To-Check requires Black not to move at all, unless he can check, in which case he must do so. White can only check if Black can annul it with a checking move, unless it is checkmate.

No-Capture-Chess is the same as Neither-Side-May-Capture, except that capturing is allowed to save the king. Must-Capture-Chess requires both sides to capture if it is possible.

Progressive chess (Scotch chess) begins with the first player making a single move. The second player then makes two moves. The first player then makes three moves, and so on. Check may only be given on the final move of a series. It must be annulled on the first move of the next player's series or it is mate.

Billiard chess allows pieces to bounce off the edges as in a game of pool, capturing up to four men in the process on the edge of the board and a fifth, after the final rebound, in the center. Kings rebound one square as a rook or bishop. Bishops may rebound from a corner square back along the diagonal. Knights may not rebound from a corner square. Pawns rebound on an edge capture on a one-space diagonal.

For similar mathematical play refer to Chess; Chess Problems.

BIBLIOGRAPHY

Dickens, Anthony. *A Guide to Fairy Chess*. New York: Dover, 1971.
Kraitchik, Maurice. *Mathematical Recreations*. 2d rev. ed. New York: Dover, 1953.
Murray, H. J. R. *A History of Chess*. 1913. Reprint. London: Oxford University Press, 1962.

**FALLACIES** are defined by W. L. Schaaf, "Number Games and Other Mathematical Recreations," as "improper reasoning leading to an unexpected result that is patently false or absurd."

If the fallacy is intentional it is called a sophism, as in Zeno's paradoxes (*see* Zeno's Paradoxes). If a fallacy leads innocently to a correct answer, it is called an illegal operation, a howler, a lucky boner, or making the right mistake (*see* Illegal Operation).

The following fallacy appeared in *Current Literature* 2 (April 1889): 349 (no author given), and was reprinted in Martin Gardner's *The "Scientific American" Book of Mathematical Puzzles and Diversions:*

Ten weary, footsore travelers,
  All in a woeful plight,
Sought shelter at a wayside inn
  One dark and stormy night.

"Nine rooms, no more," the landlord said,
  "Have I to offer you.
To each of eight a single bed,
  But the ninth must serve for two."

A din arose. The troubled host
  Could only scratch his head,
For of those tired men no two
  Would occupy one bed.

The puzzled host was soon at ease—
  He was a clever man—
And so to please his guests devised
  This most ingenious plan.

In room marked A two men were placed,
  The third was lodged in B,
The fourth to C was then assigned,
  The fifth retired to D.

In E the sixth he tucked away,
  In F the seventh man,
The eighth and ninth in G and H,
  And then to A he ran,

Wherein the host, as I have said,
  Had laid two travelers by;
Then taking one—the tenth and last—
  He lodged him safe in I.

Nine single rooms—a room for each—
  Were made to serve for ten;
And this it is that puzzles me
  And many wiser men.

Maurice Kraitchik, *Mathematical Recreations,* includes the following:

Mr. Smith meets Mr. Jones, and each of them is wearing a new necktie he received as a present. They begin to argue over which of them received the more expensive tie and agree to settle the argument by visiting the store where the ties were purchased and checking their value. They decide that the man who *wins* (has the most expensive tie) must give his tie to the loser.

Both men reason as follows: I have a fifty percent chance of winning. If I win I will lose my tie. If I lose I gain a more expensive tie. Therefore, the contest is clearly in my advantage.

According to W. L. Schaaf, an algebraic fallacy usually involves a violation of one of the following assumptions:

1. $a = b$, then $a/k = b/k$, if $k \neq 0$.
2. If $a > b$, then $ka > kb$, if $k$ is positive.
3. If $a$ is not a negative, then $\sqrt{a^2} = +a$.

Schaaf gives the following examples:

1. Solve: $6x - 18 = 4x - 12$
   Factoring: $3(2x - 6) = 2(2x - 6)$
   Dividing by $(2x - 6) : 3 = 2$.
2. Since $\sqrt{+1/-1} = \sqrt{-1/+1}$, then $\sqrt{+1/-1} = \sqrt{-1/+1}$, and so $(\sqrt{+1})(\sqrt{\neq 1}) = (\sqrt{-1})(\sqrt{-1})$, thus $+1 = -1$.
3. Given two positive numbers, $a$ and $b$:

   Then,                              Also,

   $a > -b$                           $b > -a$

$$b > -b \qquad\qquad a > -a$$

Multiplying: $ab > b^2$            $ab > a^2$

Or $a > b$                  $b > a.$

Thus, $a$ is both greater than and less than $b$.

For related mathematical recreations refer to Paradoxes of the Infinite; Zeno's Paradoxes.

BIBLIOGRAPHY

Gardner, Martin. *Aha! Gotcha: Paradoxes to Puzzle and Delight.* San Francisco: W. H. Freeman and Co., 1982.

————. *The "Scientific American" Book of Mathematical Puzzles and Diversions.* New York: Simon & Schuster, 1956.

Kraitchik, Maurice. *Mathematical Recreations.* 2d rev. ed. New York: Dover, 1953.

Schaaf, William L. "Number Games and Other Mathematical Recreations." In *Macropaedia: Encyclopaedia Britannica,* 1985.

**FAREY SERIES.** See NUMBER PATTERNS, TRICKS, AND CURIOSITIES.

**FERMAT NUMBERS.** See NUMBER PATTERNS, TRICKS, AND CURIOSITIES.

**FERRYING PROBLEMS.** See LOGICAL PROBLEMS AND PUZZLES.

**FIBONACCI NUMBERS OR SERIES.** See NUMBER PATTERNS, TRICKS, AND CURIOSITIES.

**THE FIFTEEN PUZZLE,** also called the Boss, or the Fourteen/Fifteen Puzzle, Jeu de Taquin, and Diablotin, is the most famous of the sliding-block puzzles, which are in turn the most popular of the various permutational games and puzzles.

The original puzzle consisted of a shallow square tray or box that held fifteen square counters numbered one through fifteen and an equal blank space in the lower right-hand corner. The object was to slide the fifteen numbered counters (without lifting them from the box) about the tray until they were in the desired order. To make the problem seem as simple as possible, all of the numbered squares were already in serial order, except for numbers fourteen and fifteen, which were reversed:

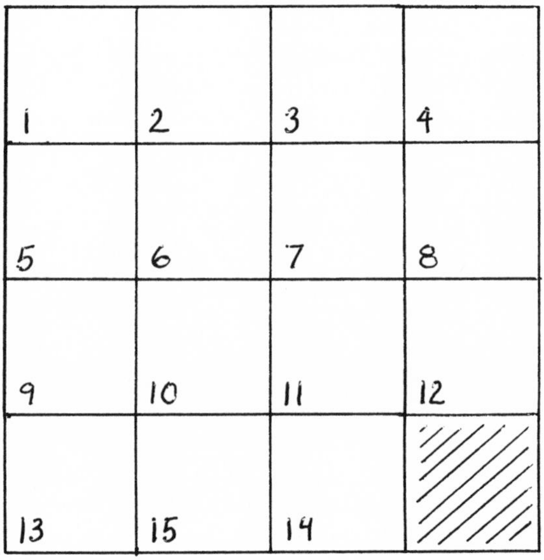

| 1 | 2 | 3 | 4 |
| 5 | 6 | 7 | 8 |
| 9 | 10 | 11 | 12 |
| 13 | 15 | 14 |  |

Samuel Loyd thought up the puzzle in the 1870s, and it was immediately popular in America and throughout Europe. Later, he was to write in his *Cyclopedia of Puzzles:*

The older inhabitants of Puzzleland will remember how in the early seventies I drove the entire world crazy over a little box of movable blocks which became known as the 14-15 Puzzle. People became infatuated with the puzzle, and ludicrous tales are told of shop-keepers who neglected to open their stores; of a distinguished clergyman who stood under a street lamp all through a wintry night trying to recall the way he had performed the feat. . . . A famous Baltimore editor tells how he went for his noon lunch and was discovered by his frantic staff long past midnight pushing little pieces of pie around on a plate! [Quoted in Martin Gardner, "Mathematical Games: The Hypnotic Fascination of Sliding-Block Puzzles."]

Loyd offered $1,000 to anyone who could solve the puzzle. Thousands responded, but no one could offer adequate proof. The reason is simple: It cannot be solved. In 1879 the *American Journal of Mathematics* published an article by William Woolsey Johnson which demonstrated why this was the case. It was immediately followed in the journal by an article by William E. Story expanding on Johnson's demonstration to show that what he calls "natural" and "reversed natural'' orders cannot be converted into one another. The editors of the journal included the following footnote:

The "15" puzzle for the last few weeks has been prominently before the American public, and may safely be said to have engaged the attention of nine out of ten persons of both sexes and of all ages and conditions of the community. But this would not have weighed with the editors to induce them to insert articles upon such a subject in the American Journal of Mathematics, but for the fact that the principle of the game has its root in what all mathematicians of the present day are aware constitutes the most subtle and characteristic conception of modern algebra, viz: the law of dichotomy applicable to the separation of the terms of every complete system of permutations into two natural and indefeasible groups, a law of the inner world of thought, which may be said to prefigure the polar relation of left and right-handed screws, or of objects in space and their reflexions in a mirror. Accordingly the editors have thought that they would be doing no disservice to their science, but rather promoting its interests by exhibiting this *a priori* polar law under a concrete form, through the medium of a game which has taken so strong a hold upon the thought of the country that it may almost be said to have risen to the importance of a national institution. Whoever has made himself master of it may fairly be said to have taken his first lesson in the theory of determinants.

It may be mentioned as a parallel case that Sir William Rowan Hamilton invented, and Jacques & Co., the purveyors of toys and conjuring tricks, in London (from whom it may possibly still be procured), sold a game called the "Eikosion" game, for illustrating certain consequences of the method of quaternions.

What all this means is that, because of what permutation mathematicians call "parity," only one-half of the 20 trillion possible positions the pieces can assume in the Fifteen Puzzle are possible from any beginning position.

Briefly, the reasoning is as follows. No matter what path a counter takes to reach the lower right corner, it must pass through an even number of boxes. If all the numbers are in serial order, no number precedes any number smaller than itself. In any other arrangement one or more numbers will precede numbers smaller than themselves. Each time this happens, it is called an inversion. For example, in the sequence 7, 2, 3, 1, 8, seven precedes three numbers smaller than itself, two precedes one number smaller than itself, and three precedes one number smaller than itself, making a total of five inversions. If the total number of inversions is even, the puzzle can be solved. If the total number of inversions is odd, the puzzle cannot be solved.

Loyd's puzzle contained only one inversion and could not be solved. He probably realized this when he published his reward offer. No one doubted his mathematical genius. He had published his first chess problem at the age of fourteen, was made problem editor of *Chess Monthly* at sixteen, and began writing a weekly chess page for *Scientific American Supplement* in 1877. By the time he died in 1911, he was to have established himself as America's un-disputed puzzle king for over half a century.

The Fifteen Puzzle gained renewed popularity in the 1940s and can be bought, along with a similar puzzle, the 31 Puzzle, in many notion and toy stores today. In both cases, today's versions normally contain the correct polarity for a solution.

In the early 1980s, these two-dimensional puzzles found renewed life in three-dimensional models beginning with the extremely popular Rubik's Cube.

Martin Gardner includes a number of variations of sliding-block puzzles that have become popular over the past one hundred years. According to Gardner, the earliest and most popular of these (though its origin is unknown) can be found in the puzzle collection of the late Lester Grimes of New Rochelle, N.Y., under the title of the Pennant Puzzle, copyrighted in 1909 by L. W. Hardy and produced by the O.K. Novelty Company in Chicago.

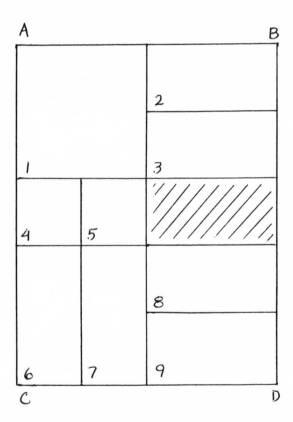

In this early version, the pieces each had the name of a city, and the large square (#1) had the name of the home team. The object was to move the home team (#1) from the upper left corner to the lower left corner, thus symbolically moving it into first place. In 1926 a wooden version was marketed under the name Dad's Puzzler. In the 1960s one with a large piano on the #1 square was marketed under the name Moving Day Puzzle, and another with wooden pieces containing magnets was marketed under the name Magnetic Square Puzzle.

According to Gardner, it is probably not possible to slide the large square at corner A to corner C in fewer than 59 moves.

The L'Ane Rouge (Red Donkey), also called the Intrique, Mov-it Puzzle, and Hako, is a popular sliding-block puzzle in France formed by cutting one of the two-by-one rectangles in half to make two unit squares, resulting in a ten-piece puzzle. The object is to move the large square with the donkey's picture on it to the bottom of the square so that it can be slid out of the box.

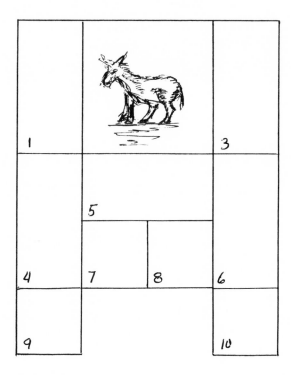

When the Dionne quintuplets were born in 1934, the Line Up or Quinties Puzzle was created (according to Gardner, the box bears the imprint of the Embossing Company of Albany, N.Y., and gives credit for its creation to Richard W. Fatiguant). The object is to transform the pattern in the first illustration (the circles represent the five quintuplets) to the pattern in the second illustration. Gardner has found a 30-move solution.

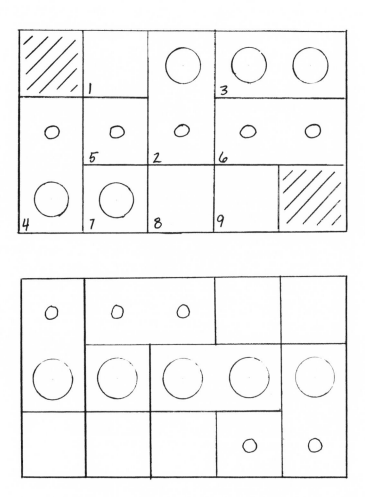

A more complex form of sliding puzzle was created by introducing non-rectangular pieces. In 1927 Charles L. A. Diamond of Newburgh, N.Y., obtained patent No. 1,633-397 for Ma's Puzzle. It was manufactured by the Standard Trailer Company of Cambridge Springs, Pennsylvania. Piece No. 2 was labeled "Ma," No. 5 "My Boy," and the other seven "No Work," "Danger," "Broke," "Worry," "Trouble," "Homesick," and "Ill." The object is to unite "Ma" with "My Boy" to form a single rectangle in the upper right-hand corner of the box. Gardner has found a 32-move solution.

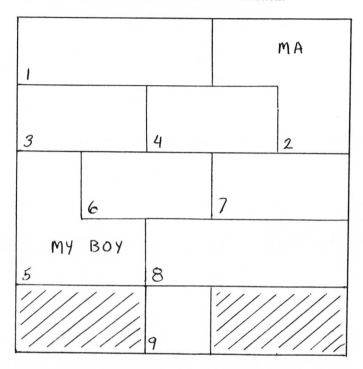

Sherley Ellis Stotts, a blind piano tuner, has invented a series of Tiger sliding-block puzzles based on visuals of the square of a polynomial:

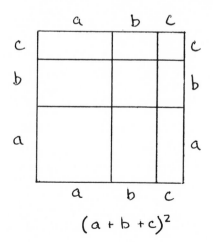

$$(a + b + c)^2$$

The puzzle based on the above polynomial starts as follows:

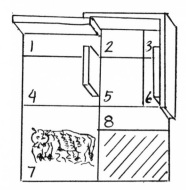

The object is to slide the pieces until the tiger square is located in the upper right-hand corner and completely surrounded by a square fence. Pieces may be rotated in the opening, if it is possible to do so without lifting them. Stotts has accomplished this in 49 moves. According to Gardner, Stotts' other Tiger sliding-block puzzles are even more difficult than this one.

For similar mathematical puzzles refer to Chess; Chess Problems; Fairy Chess; Graphs and Networks; The Tower of Hanoi.

BIBLIOGRAPHY

Ball, W. W. Rouse. *Mathematical Recreations and Essays*. 1892; Rev. by H. S. M. Coxeter. London: Macmillan and Co., 1939.
Gardner, Martin. "Mathematical Games: The Hypnotic Fascination of Sliding-Block Puzzles." *Scientific American* 210 (Feb. 1964): 122–130.
_____. The "Scientific American" Book of Mathematical Puzzles and Diversions. New York: Simon & Schuster, 1956.
Johnson, William Woolsey. "Notes on the '15' Puzzle, I." *American Journal of Mathematics* 2 (1879): 397–399.
Kraitchik, Maurice. *Mathematical Recreations*. 2d rev. ed. New York: Dover, 1953.
Northrop, Eugene P. *Riddles in Mathematics: A Book of Paradoxes*. New York: D. Van Nostrand Co., 1944.
Schaaf, W. L. "Number Games and Other Mathematical Recreations." In *Macropaedia: Encyclopaedia Britannica*, 1985.
Spitznagel, Edward L., Jr. "A New Look at the Fifteen Puzzle." *Mathematics Magazine* 40 (Sept. 1967): 171–174.
Steinhaus, H. *Mathematical Snapshots*. 3d rev. ed. New York: Oxford University Press, 1969.
Story, William E. "Notes on the '15' Puzzle, II." *American Journal of Mathematics* 2 (1879): 399–404.

**FIFTH POSTULATE.** See EUCLID'S THEORY OF PARALLELS.

**FIGURATE NUMBERS.** See NUMBER PATTERNS, TRICKS, AND CURIOSITIES.

**FIVE UP.** See DOMINOES.

**FLEXAGONS.** See GEOMETRIC PROBLEMS AND PUZZLES.

**FOOL'S MATE.** See CHESS PROBLEMS.

**FOUR FIELD KONO** is a board game.
  Two players start with eight counters, placed on the board as follows:

The object is to capture or block all of the opponent's markers. Each player moves in turn. To capture an opponent's marker it is necessary to jump over one of the markers of the player making the move and land directly on top of the opposing player's marker. Captured markers are removed from the board. When not capturing, a piece may only be moved along one of the connecting lines one space at a time.
  For similar recreations refer to Checkers; Chess; Topology.

BIBLIOGRAPHY

Brandreth, Gyles. *The World's Best Indoor Games*. New York: Pantheon Books, 1981.

**FOURTEEN/FIFTEEN PUZZLE.** See THE FIFTEEN PUZZLE.

**FOX AND GOOSE** is a board game.
  The board is set up as follows:

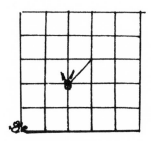

A dime is placed over the fox and a penny over the goose. One player moves the fox, the other the goose. A move consists of sliding the coin from one dot to an adjacent dot along a black line. The object is for the fox to capture the goose by moving onto the spot occupied by the goose. The goose tries to prevent capture. If the fox captures the goose in ten moves or less, the fox wins; otherwise, the goose wins.

For similar recreations refer to Topology.

BIBLIOGRAPHY

Gardner, Martin. *Mathematical Puzzles*. New York: Thomas Y. Crowell Co., 1961.

**FRANKLIN SQUARES.** See MAGIC SQUARES.

**FRENCH DRAW.** See DOMINOES.

# G

**GEMATRIA,** a form of cryptogram, is a superstitious use of letters to represent simultaneously words and numbers, thus allowing words to take on special significance or symbolic meanings.

A combination of gematria and Pythagorean number symbolism served as the basis for the number magic of the medieval Cabala and allowed secret readings of the Hebrew Bible.

A famous example from the New Testament is the number of the Beast (666) found in the thirteenth chapter of Revelation. Many attempts have been made to discover the name of the Beast from this number, but no generally accepted solution has come forth. It is possible that the numbers are meant to represent Nero and Caesar, for the Hebrew characters of these names give a numerical total of 666. Beasting the Man is the giving to that man a numerical value of 666 (generally by assigning numbers to the letters of his name or birthday). Among the famous names that have been Beasted are Martin Luther, Pope Leo X, and Napoleon.

Heliodorus, a Greek philosopher, proved from gematria that ''Nile'' means ''year,'' because the name written in Greek capital letters is $N$ *(50), E (5), I (10),* $\Lambda$ *(30), O (70),* and $\Sigma$ *(200),* which equals *365.*

Pairs of numbers are called amicable or sympathetic if the sum of the divisors of each equals the other number, e.g.:

$284 = 110 + 55 + 44 + 22 + 20 + 11 + 10 + 5 + 4 + 2 + 1$ (the divisors of 220), and
$220 = 142 + 71 + 4 + 2 + 1$ (the divisors of 284).

People whose names result in amicable numbers are said to make good friends.

In Hebrew, consonants are used as numerical signs, but by adding vowels they can be read as words, e.g., *JHVH* can be read as either Yahweh or Jehovah.

For more discussion on attaching mystical numerical significance to letters, refer to Numerology. For mathematical games and puzzles based on the combination of numbers and letters, refer to ABC Words; ACE Words; Centurion; Numwords.

BIBLIOGRAPHY

Bell, E. T. *Numerology: The Magic of Numbers.* New York: United Book Guild, 1945.
Bourguignon, Erika. "Numerology." In *Encyclopaedia Americana,* International Ed., 1985.
Bowers, Henry, and Joan E. Bowers. *Arithmetical Excursions: An Enrichment of Elementary Mathematics.* New York: Dover, 1961.
Friend, J. Newton. *More Numbers: Fun and Facts.* New York: Charles Scribner's Sons, 1961.
Frohlichstein, Jack. *Mathematical Fun, Games and Puzzles.* New York: Dover, 1967.
Gardner, Martin. *The Incredible Dr. Matrix.* New York: Charles Scribner's Sons, 1976.
––––––. *Mathematical Carnival.* New York: Alfred A. Knopf, 1965.
––––––. *The Numerology of Dr. Matrix.* New York: Simon and Schuster, 1967.

**GEOMETRIC DISSECTIONS,** also called dissection puzzles, involve the cutting up of planes (and figures of higher dimensions) to make other figures.

According to Martin Gardner, "Mathematical Games: Wherein Geometrical Figures Are Dissected to Make Other Figures," the first systematic treatment of the subject seems to have been made by Abul Wefa, a tenth-century Persian mathematician. Only fragments of his book still survive. Nevertheless, they do include a number of interesting dissections. The following example (included in Gardner's article) shows Abul Wefa's nine-piece solution to how to dissect three identical squares into pieces that can be recombined to form one single square:

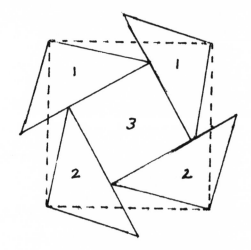

Squares #1 and #2 have been divided diagonally and placed around square #3 as shown. Four more cuts along the dotted lines indicate how the task is completed.

Strong interest in geometric dissections did not begin, however, until the twentieth century, when such men as Henry E. Dudeney (1857–1931) began attempting solutions with the added challenge of completing the various problems *with the fewest number of pieces*. Dudeney's solution to the problem of Wefa included above (also included in Gardner's article) still stands as the one with the fewest pieces (six):

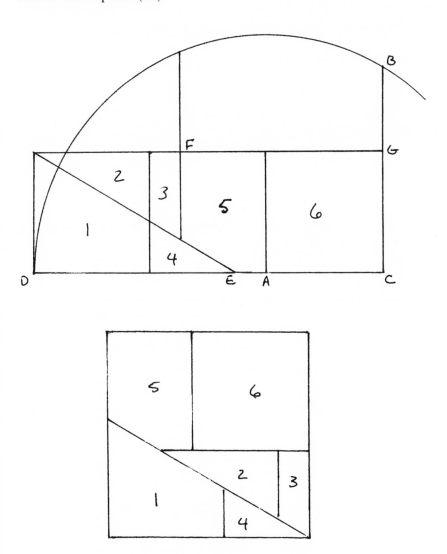

The circle has its center at *A. BC = DE = FG.*

Harry Lindgren, the Examiner of Patents at the Australian Patent Office in Canberra, Australia, is far and away the most important person in the field. Though he has dealt with solid dissections and other aspects of geometrical dissections, most of his work concerns the task of reforming polygons into other polygons of the same area in the minimum number of moves, and he holds the record for the smallest number of pieces for numerous dissections.

Though there is no general procedure applicable to all geometric dissections, Lindgren has discussed his methods at some length in his book, *Geometric Dissections,* where such techniques as the parallelogram slide, the quadrilateral slide, and strip dissections are revealed. The following is how a strip dissection works:

Take a polygon, say, a Latin cross:

Cut the Latin cross so that its pieces fit together to form a potentially endless strip with parallel top and bottom lines:

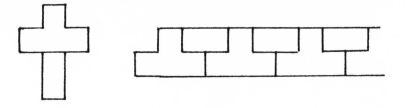

Do the same with whatever polygon the first polygon is to be changed into, say, a square:

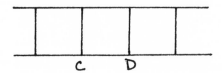

Superpose the second figure over the first, making sure the congruent points
(e.g., *C* and *D*) lie on the parallel edge lines of the first strip:

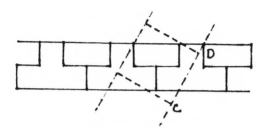

Now slide the superposed figure along the strip until the number of crossing
points is reduced to a minimum:

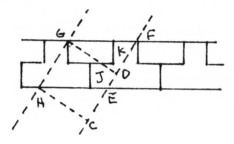

If pieces *J* and *K* in the *EFGH* parallelogram are moved to the left so that *EF* lies
along *GH*, it is easy to see how the figure can be changed into the following Latin
cross:

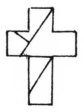

It is also possible to rearrange these same pieces into a square:

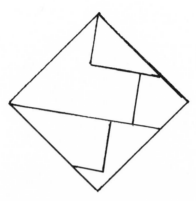

The dissection of a square into two smaller, unequal squares has long been used as a proof of the Pythagorean theorem (i.e., the square of the hypotenuse in a right triangle is equal to the sum of the squares of the other two sides—$c^2 = a^2 + b^2$) in the following triangle:

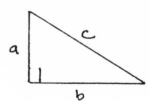

By employing geometric dissection this theorem can be demonstrated visually. In the following figure, the square whose side is equal to the hypotenuse of the right triangle has been geometrically divided up and reformed to make up the squares whose sides are equal to the other two sides of the triangle:

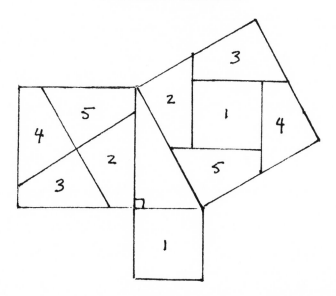

The least complex geometric dissection is that of a 1 × 4 rectangle into a square:

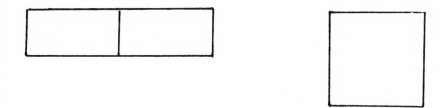

If the ratio of the length of a rectangle to the width is not 1 to 4, a straight-cut two-piece dissection is not possible. It is possible to use a step-cut technique for some other rectangles (e.g., a 4 × 9 one) and only use two pieces:

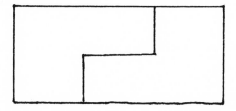

The above dissection has a "two-step" look about it and can be referred to as a two-step cut. A $9 \times 16$ rectangle can be converted with three steps, a $16 \times 25$ with four, a $20 \times 20$ with five, and so on. The formula for this is: an $n$-step, two-piece dissection of a rectangle to a square is possible if the dimensions are in the ratio of $(n + 1)^2$ to $n^2$ where the ratio is equal to or less than 4 to 1. As $n$ approaches infinity, the steps become so small as to approach a straight diagonal:

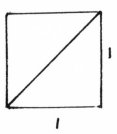

Another type of geometric dissection problem involves squared rectangles and "squaring the square." A squared rectangle is a rectangle that can be dissected into a finite number of squares. If none of these squares are equal, the squared rectangle is said to be perfect.

Squaring the square is dissecting a square into smaller squares, no two of the same size. It was thought to be impossible until it was finally accomplished in 1939.

For similar recreations refer to Graphs and Networks; Mazes; and Labyrinths; Tangrams.

BIBLIOGRAPHY

Gardner, Martin. "Mathematical Games: Wherein Geometrical Figures Are Dissected to Make Other Figures." *Scientific American* 25, no. 5 (November 1961): 158–170.
_____. *The Second "Scientific American" Book of Mathematical Puzzles and Diversions.* New York: Simon & Schuster, 1961.
Kinnard, Clark, ed. *Encyclopedia of Puzzles and Pastimes.* New York: Citadel Press, 1946.
Lindgren, Harry. *Geometric Dissections.* New York: D. Van Nostrand Co., 1964.
MacMahon, P. A. *New Mathematical Pastimes.* London: Cambridge University Press, 1930.
Madachy, Joseph S. *Mathematics on Vacation.* New York: Charles Scribner's Sons, 1966.
Mott-Smith, Geoffrey. *Mathematical Puzzles for Beginners and Enthusiasts.* Philadelphia: Blakiston Co., 1946.
Schaaf, W. L. "Number Games and Other Mathematical Recreations." In *Macropaedia: Encyclopaedia Britannica,* 1985.

**GEOMETRIC PROBLEMS AND PUZZLES,** also called space puzzles, are those mathematical activities involving properties, measurements, and relationships of points, lines, angles, and figures in space.

*Continuous line* or *trace puzzles* are those puzzles that challenge the player to draw a line without lifting his pencil from the paper through a certain configuration (usually a series of dots); e.g., draw the following figure in one continuous line without retracing any part of it:

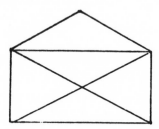

The solution:

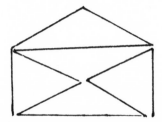

Chess problems (*see* Chess Problems) involving tours of the board are an activity of this type. Continuous line or trace puzzles are a form of graph and network recreation (*see* Graphs and Networks).

*Origami* or *paper folding puzzles* are those problems in mathematics concerning the properties of folded paper. A square can be formed by folding paper as follows:

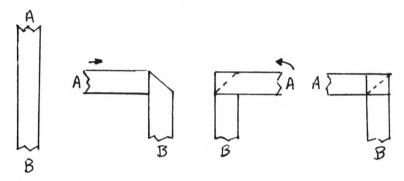

A regular heptagon can be formed as follows:

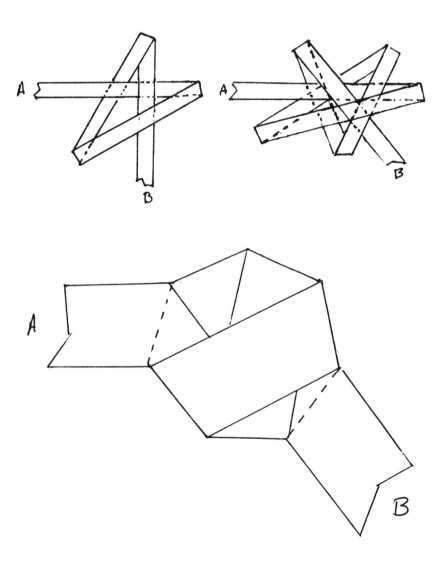

*Flexagons* are polygons constructed from a strip of paper in such a way that the figure possesses the property of changing its faces when it is "flexed." Flexagons were originated by Arthur H. Stone in 1939. Apparently, as an English student taking postgraduate study in mathematics at Princeton University, he found it necessary to trim strips from his American notebook paper so that it would fit in the binder he had brought from Britain. Upon playing with the leftover strips, he produced the following flexagon:

a.

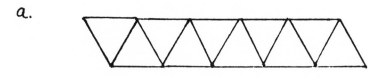

b.

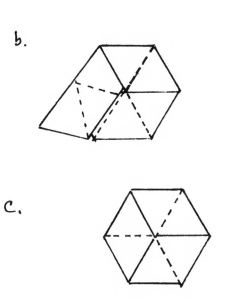

c.

A strip of paper is cut and equilateral triangles drawn on it as in part a. Of the above figure three folds are made away from the folder, and the final triangle is pasted onto the first triangle, resulting in part c. The flexagon may then be folded into a shape like that of the feathers of an arrow or dart and reopened at the center to reveal an entirely new face.

Depending on the number of equilateral triangles on the original strip, increasingly complex flexagons are possible. The above flexagon with ten triangles is called a trihexaflexagon. The day after Stone came up with his flexagon, he produced a hexagonal model that could be folded with six faces instead of only three. He showed his paper models to his friends—Bryant Tuckerman, Richard P. Feynman, and John W. Turkey—and the "Flexagon Committee" was formed. They named their models *hexaflexagons* (*hexa* for the hexagonal form, and *flexagon* for their ability to flex). Stone's second model, the one with six faces, was called a *hexahexaflexagon*.

A hexaflexagon is flexed by pinching together two of the triangles. Tuckerman discovered that the easiest way to flex out all of the faces was to keep flexing at the same corner until it refused to open and then shift to an adjacent corner (the procedure is appropriately called the Tuckerman Traverse). The committee discovered that it is possible to make nine, twelve, fifteen, or more faces. Tuckerman made a working model with forty-eight. In addition, Tuckerman found that it was possible to produce tetrahexaflexagons (four faces) and pentahexaflexagons by employing a sawtooth (zigzag) pattern instead of the straight edges.

*Geometric dissections* and *dividing the plane puzzles* involve the cutting up and manipulating of space (*see* Geometric Dissections). Here are a few examples:

How can a cake be cut into eight equal pieces with only three cuts? The solution: Two vertical cuts are made at right angles to each other along the diameters and a horizontal cut is made horizontally through the middle of the cake.

How can a separate pen be made for each sheep in the following enclosure by building only two more square enclosures?

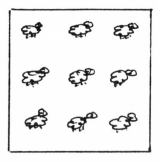

The solution:

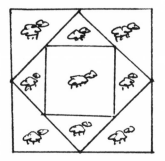

How can a separate pen be made for each sheep in the following enclosure by drawing only three lines?

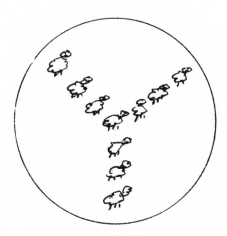

The solution:

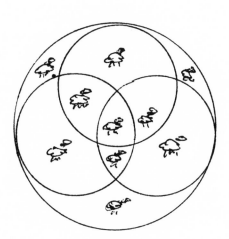

How many triangles are in this figure?

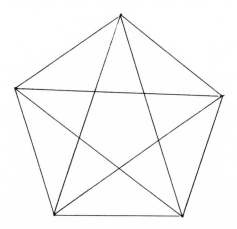

The solution: 35.

How many squares are in this figure?

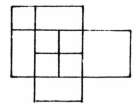

The solution: 11.

For more play involving geometry refer to Checkers; Chess; Chess Problems; Dominoes; The Duplication of the Cube; Fairy Chess; Geometric Dissections; Graphs and Networks; Match Problems, Puzzles and Games; Mazes and Labyrinths; The Möbius Strip; Paradoxes of the Infinite; Squaring the Circle; Tangrams; Topology; The Trisection of an Angle; Zeno's Paradoxes.

BIBLIOGRAPHY

Ball, W. W. Rouse. *Mathematical Recreations and Essays*. 1892. Rev. by H. S. M. Coxeter. London: Macmillan and Co., 1939.
_____. *A Short Account of the History of Mathematics*. London: Macmillan and Co., 1888.

Behnke, H., F. Bachmann, K. Fladt, and H. Kunle. *Fundamentals of Mathematics. Vol. 2: Geometry.* Cambridge, Mass.: MIT Press, 1974.

Bell, E. T. *The Development of Mathematics.* New York: McGraw-Hill, 1945.

Dudeney, Henry Ernest. *Amusements in Mathematics.* 1917. Reprint. New York: Dover, 1958.

_____. *536 Puzzles and Curious Problems.* Reprint. Ed. by Martin Gardner. New York: Dover, 1967.

Fixx, James F. *More Games for the Super-Intelligent.* New York: Doubleday, 1976.

Frohlichstein, Jack. *Mathematical Fun, Games and Puzzles.* New York: Dover, 1967.

Gardner, Martin. *Mathematical Puzzles.* New York: Thomas Y. Crowell Co., 1961.

_____. *The "Scientific American" Book of Mathematical Puzzles and Diversions.* New York: Simon & Schuster, 1959.

_____. *The Second "Scientific American" Book of Mathematical Puzzles and Diversions.* New York: Simon & Schuster, 1961.

Kinnard, Clark, ed. *Encyclopedia of Puzzles and Pastimes.* New York: Citadel Press, 1946.

Kraitchik, Maurice. *Mathematical Recreations.* 2d ed., rev. New York: Dover, 1953.

Lindgren, Harry. *Geometric Dissections.* New York: D. Van Nostrand Co., 1964.

MacMahon, P. A. *New Mathematical Pastimes.* London: Cambridge University Press, 1930.

Madachy, Joseph S. *Mathematics on Vacation.* New York: Charles Scribner's Sons, 1966.

Schaaf, W. L. "Number Games and Other Mathematical Recreations." In *Macropaedia: Encyclopaedia Britannica,* 1985.

Trigg, Charles W. *Mathematical Quickies.* New York: McGraw-Hill, 1967.

**GEOMETRIC PROGRESSION.** See NUMBER PATTERNS, TRICKS, AND CURIOSITIES.

**GNOMONS.** See NUMBER PATTERNS, TRICKS, AND CURIOSITIES.

**GOLDEN RATIO.** See NUMBER PATTERNS, TRICKS, AND CURIOSITIES.

**GOLDWYNNER.** See LOGICAL PARADOXES.

**GOLOMB'S GAME.** See DOMINOES.

**GRAPH AND CHOPPER.** See NIM.

**GRAPHS AND NETWORKS** are geometric recreations that involve connecting or rearranging points on a plane or in space. The term "graph" is not used here to refer to the curves of analytic geometry and function theory.

Dénes König made the first systematic study of graphs in the 1930s; Claude

Berge's book on graph theory was published in England in 1962; and Wystein Ore's *Graphs and Their Uses*, an introduction to graphs, was published in 1963. Today, graphs and graph theory comprise an important field of mathematics.

Planar graphs are graphs of points (vertices) connected by lines (edges) so that it is possible to draw the graph on a plane without any pair of edges intersecting. If each pair of vertices is connected by an edge, the graph is called a complete graph. In the following illustrations, a. is a complete graph, b. is a nonplanar graph, and c. is an equivalent or isomorphic planar graph of b.:

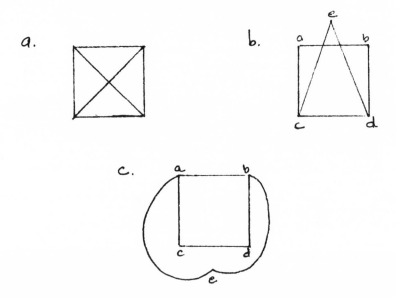

In the illustrations above, a. is a planar graph in addition to being a complete graph (the crossing point at the center is not a vertex—think of the lines as passing under and over each other).

Here is a graph puzzle involving three wells. In the following diagram, *A, B,* and *C* represent three neighbors' houses, and *X, Y,* and *Z* represent three wells:

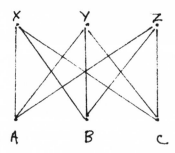

The object is to find paths leading from each house to each well so that none of the paths crosses another path. It is impossible, and the proof is based on the Jordan curve (*see* Topology). This same puzzle was given the form of connecting three houses and their sources of gas, water, and electricity by Henry Ernest Dudeney in 1917 and became known as the *utilities puzzle*.

A complete graph can be planar only if it has four or fewer points.

An interesting form of graph puzzle is the *continuous line* or *trace puzzle* (*see* Geometric Problems and Puzzles). The object is to draw a given planar graph in one continuous line without taking the pencil from the paper or retracing any lines. If it can be done, returning to the vertex from where it started, it is said to be an *Euler graph*, named after Leonhard Euler (eighteenth century), who solved what has been called the *Konigsberg* (now Kaliningrad) *Bridge problem*. The problem was to determine if it was possible to walk across the seven bridges that span the Pregel (Pregolya) River as follows:

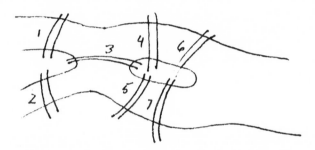

and return to the starting position without crossing any bridge more than once. Euler's solution (demonstrating that it is impossible) was the first to deal with modern graph theory. He proved the following principles: (1) The number of even points (those in which an even number of points meet) is irrelevant. (2) The number of odd points is always even. (3) If there are no odd points the network is unicursal (can be traversed in a closed loop). (4) If there are exactly two odd points, it is possible to start at one and finish at the other; if there are more than two odd points, the network cannot be traced in one continuous path.

Sir William Rowan Hamilton was the first to study paths that would pass through each vertex once and only once. Any route that passes through no vertex more than once is known as an arc. Any arc that returns to its starting point is called a circuit. Any circuit that passes through each vertex once and only once is called a Hamiltonian line. Hamilton (1859) came up with a puzzle that required finding a route through each point on a regular dodecahedron by following its edges once and only once. For problems of this sort refer to Chess Problems.

For additional recreations in this area refer to Geometric Problems and Puzzles.

BIBLIOGRAPHY

Gardner, Martin. *Martin Gardner's Sixth Book of Mathematical Games from "Scientific American."* San Francisco: W. H. Freeman and Co., 1963.

Mott-Smith, Geoffrey. *Mathematical Puzzles for Beginners and Enthusiasts.* Philadelphia: Blakiston Co., 1946.

Schaaf, W. L. "Number Games and Other Mathematical Recreations." In *Macropaedia: Encyclopaedia Britannica,* 1985.

Schuh, Fred. *The Master Book of Mathematical Recreations.* Trans. by F. Göbel; ed. by T. H. O'Beirne. New York: Dover, 1968.

**GREAT CHESS.** See FAIRY CHESS.

**GREGORIAN CALENDAR.** See CALENDARS.

**GRILLES.** See CRYPTARITHMS, CRYPTOGRAPHY, CONCEALMENT CIPHERS, SUBSTITUTION CIPHERS, TRANSPOSITION CIPHERS, AND CODE MACHINES.

**GROUPING PUZZLES.** See ARITHMETIC AND ALGEBRAIC PROBLEMS AND PUZZLES.

**GUESSING A SELECTED NUMBER.** See GUESSING NUMBERS.

**GUESSING NUMBERS,** also called number guessing and guessing a selected number, involves finding a number chosen by someone (providing the result of certain operations on it is known).

Here are a few of the common methods:

1. Ask a person to choose a number and triple it and say whether the result is even or odd. If the number is odd, have the person add 1 to it and divide it by 2. If it is even, simply have the person divide it by 2. Then have the person triple the result and tell you how many times 9 goes into the result (disregarding any remainder). If $x$ stands for this result, the $2x$ or $2x + 1$ equals the number selected (depending on whether the result of his first operation was even or odd).

In mathematical formula, the operation works as follows: If the number was even, say, $2x$, then the first operation is $2x \cdot 3 = 6x$; the second operation is $6x/2 = 3x$; the third operation is $3x \cdot 3 = 9x$; and the final operation is $9x/9 = x$. If the number was odd, say, $2x + 1$, then the first operation is $(2x + 1)3 = 6x + 3$; the second operation is $(6x + 3 + 1)/2 = 3x + 2$; the third operation is $(3x + 2)3 = 9x + 6$; and the final operation is $(9x + 6)/9 = x$, with a remainder of 6 to discard.

2. Ask someone to think of a number, multiple it by 5, add 6 to the product, multiply this by 4, add 9 to the product, and multiply by 5. This produces a

number of three digits or more. Cross off the final two digits. Subtract 1 from the number left. The number remaining is the number the person selected.

The mathematical formula is $[(5x + 6)4 + 9]5 = 100x + 165; x = [100x + 165 (-65)/100] - 1$.

3. Ask someone to think of a number, multiply it by any number you wish *(a)*, divide the product by any number you wish *(b)*, multiply the result by *(c)*, and divide the result by *(d)*. Then have him divide this by the number selected originally and add to this the number originally selected. Ask him for his result and subtract *ac/bd* from it. The remainder will be the original number.

If *n* equals the number selected, then *nac/bd* equals the first four operations; operation five equals *ac/bd;* and operation six results in *n* + *(ac/bd)*. You already know *ac/bd*. By subtracting this from the result mentioned, you obtain *n*.

4. Ask someone to choose a number less than 90, multiply it by 10, add any number *(a)* less than 10, divide the result by 3, and tell you the remainder *(b)*. Then have him multiply the quotient by 10 and add any number *(c)* less than 10. Then have him divide the result by 3 and tell you the remainder *(d)* and the third digit from the right of the quotient *(e)*.

If you know numbers *a, b, c, d,* and *e*, it is possible to determine the number. If the number is $9x + y$, where $x \leqq 9$ and $y \leqq 8$, and *r* is the remainder when $a - b + 3(c - d)$ is divided by 9, the result is $x = e, y = 9 - r$.

5. Have someone choose two numbers (each of them less than 10), multiply one of them by 5 and add 7, double the sum, add the other number, and give the result. By subtracting 14 from this number, the two-digit number obtained will be the digits of the two numbers selected.

6. Have several people choose different numbers from 1 through 9. Have the first person double his number, add 1, and multiply by 5. Have the next person add his number to the result, multiply by 2, add 1, and multiply by 5. Continue this through all of the people in turn. Subtract a number made up of the number of 5's equal to the number of people (e.g., if five people, 55555) from the number obtained by the final person and divide by 10. The result will be the number that contains the digits of the chosen numbers in order.

7. Have someone choose a three-digit number and designate it by the letters *a, b, c*. Have the person arrange the digits as follows—*abc, bca,* and *cab*—and divide each in turn by 11. Ask him what the remainders are *(a, b, c)*. Add these remainders as follows: $a + b, b + c, c + a$. If any of the sums is odd, increase or decrease it by 11 to obtain a number between 0 and 20. Divide each of the numbers in turn by 2 to obtain the original three digits.

8. Have someone choose a three-digit number (the first and third digits different), reverse the first and third digits, and subtract the second number from the first. Have him tell you the final digit of the number obtained. You are then able to tell him the rest of the digits.

The number will equal $100a + 10b + c$. A reversal of the digits produces *100c*

$+ 10b + a$. The difference of these two numbers equals $99(a - c)$, since $a - c$ cannot be greater than 9. Thus, the final number can only be 99, 198, 297, 396, 495, 594, 693, 792, or 891, the middle digit always equaling 9, and the outside digits totaling 9.

A method of quick addition lies at the middle of many number guessing tricks. It works as follows:

Have someone pick a number, say, 13,485. Have him pick a second number, say, 17,865. Now, pick a number so that your number added to the previous number creates a series of nines:

$17,865 + 82,134 = 99,999$

Have the person choose another number, say, 15,721. Once again, choose a number to add to this number to create a series of nines:

$15,721 + 84,278 = 99,999$

This process can be continued as long as wished. Then subtract the number of nine-groups obtained (2 in the example above), from the original number:

$13,485 - 2 = 13,483$

Place the number of nines subtracted at the beginning of this number (13,483) to obtain the answer (213,483).

If the other person chooses all nines (e.g., 99,999) for one of the numbers, you pass your turn and still count the number as a group of nines. There is no limit to the length of the list or the size of the numbers.

The process discussed above is generally referred to as *casting out nines*. Karl Friedrich Gauss (1777–1855) developed a form of it based on numerical congruence to deal with the powers of numbers called *digital roots*. It works as follows: The sum of the digits of every positive integer gives the same remainder when divided by 9 as does the integer itself when divided by 9, as does the $n$th power of the sum of these digits when divided by 9. A digital root, then, is the ultimate sum of the digits of the integer. For example, 625 has the following digital root: $6 + 2 + 5 = 13$; $1 + 3 = 4$; $4 + 0 = 4$. 4/9 leaves a remainder of 4.

This, in turn, helps in a form of digital recreation which seeks to find numbers that are equal to the $n$th power of the sums of their digits:

$81 = (8 + 1)^2 = 9^2$

Since the digital root of any integer is a single digit, only the remainder of whatever power sought of the nine digits when divided by 9 need be found.

Thus, if the first nine numbers are raised to $x$ power, the remainders obtained from the above process will be the only ones possible for solutions to the problem of finding numbers that are equal to the $n$th power of the sums of their digits:

| digit | digit$^3$ | digital root of digit$^3$ | remainder when divided by 9 |
|-------|-----------|---------------------------|------------------------------|
| 1 | 1 | 1 | 1 |
| 2 | 8 | 8 | 8 |
| 3 | 27 | 9 | 0 |
| 4 | 64 | 1 | 1 |
| 5 | 125 | 8 | 8 |
| 6 | 216 | 9 | 0 |
| 7 | 343 | 1 | 1 |
| 8 | 512 | 8 | 8 |
| 9 | 729 | 9 | 0 |

Thus, if the digital root of a number cubed has a remainder of 0, 1, or 8, it *might* satisfy the conditions imposed (only $8^3$ of the first nine numbers does); if a number cubed does not have 0, 1, or 8 for a digital root, it cannot satisfy the conditions.

For similar play refer to Digital Problems; Number Patterns, Tricks, and Curiosities.

BIBLIOGRAPHY

Ball, W. W. Rouse. *Mathematical Recreations and Essays*. 1892. Rev. H. S. M. Coxeter. London: Macmillan and Co., 1939.

Gardner, Martin. *The Second "Scientific American" Book of Mathematical Puzzles and Diversions*. New York: Simon & Schuster, 1961.

Kraitchik, Maurice. *Mathematical Recreations*. 2d rev. ed. New York: Dover, 1953.

Madachy, Joseph S. *Mathematics on Vacation*. New York: Charles Scribner's Sons, 1966.

Shaaf, W. L. "Number Games and Other Mathematical Recreations." In *Macropaedia: Encyclopaedia Britannica*, 1985.

# H

**HACKENBUSH.** See NIM.

**HELPMATE.** See FAIRY CHESS.

**HETEROSQUARE.** See MAGIC SQUARES.

**HEXAFLEXAGONS.** See GEOMETRIC PROBLEMS AND PUZZLES.

**HEXAGONAL NUMBERS.** See NUMBER PATTERNS, TRICKS, AND CURIOSITIES.

**HOTEL INFINITY.** See PARADOXES OF THE INFINITE.

**HOWLER.** See ILLEGAL OPERATION.

# I

**ICOSIAN GAME.** See THE TOWER OF HANOI.

**ILLEGAL OPERATION,** also called a howler, a lucky boner, or making the right mistake, is an error introduced into a calculation or proof that leads "innocently" to a correct answer.

It is similar to a sophism, i.e., a fallacy in which an error has been knowingly committed. A sophism, in turn, is closely related to a paradox, i.e., a conclusion so unexpected that it is hard to accept even though the reasoning is valid. All of these are closely related to a fallacy, i.e., improper reasoning that leads to an unexpected result that is false or absurd.

W. L. Schaaf offers the following example of an illegal operation:

$$\frac{16}{64} = \frac{16\!\!\!/}{64\!\!\!/} = \frac{1}{4}$$

For more in-depth discussion of paradoxes and fallacies refer to Fallacies; Paradoxes of the Infinite; Zeno's Paradoxes.

BIBLIOGRAPHY

Schaaf, William L. "Number Games and Other Mathematical Recreations." In *Macropaedia: Encyclopaedia Britannica*, 1985.

**INFINITY MACHINES.** See ZENO'S PARADOXES.

**INTERNATIONAL FIXED CALENDAR.** See CALENDARS.

**INVERSE FUNCTION.** See MIRROR.

**INVERSIONS.** See MIRRORS.

**IRISH BULL.** See LOGICAL PARADOXES.

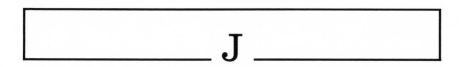

# J

**JAPANESE MAGIC CIRCLES.** See MAGIC CIRCLES.

**JEU DES DAMES.** See CHECKERS.

**JEU DE TAQUIN.** See THE FIFTEEN PUZZLE.

**JULIAN CALENDAR.** See CALENDARS.

# K

**KALEIDOSCOPE.** See MIRRORS.

**KEY WORD CIPHER.** See CRYPTARITHMS, CRYPTOGRAPHY, CONCEALMENT CIPHERS, SUBSTITUTION CIPHERS, TRANSPOSITION CIPHERS, AND CODE MACHINES.

**KLEIN BOTTLE.** See THE MÖBIUS STRIP.

**KNIGHT TOURS.** See CHESS PROBLEMS.

**KONIGSBERG BRIDGE PROBLEM.** See GRAPHS AND NETWORKS.

**KRIEGSPIEL CHESS.** See FAIRY CHESS.

# L

**LATIN SQUARES.** See MAGIC SQUARES.

**LAW OF SERIES.** See NUMBER PATTERNS, TRICKS, AND CURIOSITIES.

**LEGAL CHESS.** See FAIRY CHESS.

**LEGAL GAME.** See FAIRY CHESS.

**LEWIS CARROLL VIGENERE CIPHER.** See CRYPTARITHMS, CRYPTOGRAPHY, CONCEALMENT CIPHERS, SUBSTITUTION CIPHERS, TRANSPOSITION CIPHERS, AND CODE MACHINES.

**LIAR PARADOX.** See LOGICAL PARADOXES.

**LIZZIE BORDEN'S NIM.** See NIM.

**LOCULUS OF ARCHIMEDES.** See TANGRAMS.

**LOGICAL PARADOXES** are those self-contradictory statements that challenge the entire structure of mathematics.

In 1919, Bertrand Russell (*Introduction to Mathematical Philosophy,* p. 194), wrote:

Mathematics and logic, historically speaking, have been entirely distinct studies. Mathematics has been connected with science, logic with Greek. But both have developed in modern times: logic has become more mathematical and mathematics has become more logical. The consequence is that it has now become wholly impossible to draw a line between the two; in fact, the two are one. . . . The proof of their identity is, of course, a matter of detail: starting with premises which would be universally admitted to belong to logic, and arriving by deduction at results which as obviously belong to mathematics, we find that there is no point at which a sharp line can be drawn, with logic to the left and mathematics to the right.

In 1910 Russell and Alfred N. Whitehead published *Principia Mathematica,* an ambitious attempt to provide all of mathematics with a *logical* basis. It was greeted with controversy, and though that controversy has resulted in some major advances, it remains unresolved—a disquieting uncertainty at the very basis of mathematics.

Here are some of the paradoxes: The first comes from Russell himself and is contained in nearly every contemporary book published on mathematical paradoxes. It is generally titled "The Barber's Paradox":

A barber posts a sign on his window: "I shave all those men in town and only those men who do not shave themselves." However, when it comes time for the barber to shave himself, he is in a quandary. If he shaves himself, then he belongs to those men who shave themselves. If someone else shaves him, then he is a man who does not shave himself. His sign states that he shaves all those men and only those men who do not shave themselves.

The following is considered by many to be the first recorded logical paradox: Epimenides, a Cretan of the sixth century B.C., is reputed to have said, "All Cretans are liars." Assuming that he meant by this that all Cretans *always* lie, then he has left the world with a logical paradox. St. Paul refers to this statement in his epistle to Titus in the New Testament:

One of themselves, even a prophet of their own, said, the Cretans are always liars, evil beasts, slow bellies.
This witness is true. . . . [Titus 1:12–13]

P. E. B. Jourdain, an English mathematician, offered the following paradox in 1913:

"The sentence on the other side of this card is TRUE."
On the other side of the card, the sentence read:
"The sentence on the other side of this card is FALSE."

Martin Gardner includes a number of variations on this type of paradox (it is often referred to as the liar paradox) in *Aha! Gotcha: Paradoxes to Puzzle and Delight.* In its simplest form it goes: "This sentence is false."

According to Gardner, Bertrand Russell made the following comment about George Edward Moore only lying once in his life. When someone asked Moore if he always told the truth, he thought for a moment and said, ''No!''

There are numerous one-liners involving this type of paradox. Here are a few:

Nothing is absolute!
All rules have exceptions!
Don't read this!
George Bernard Shaw's statement: ''The only Golden Rule is that there are no rules.''
Groucho Marx' statement: ''I refuse to join any club willing to have me as a member.''

Eugene P. Northrop includes the following paradox, said to have come from Protagoras (fifth century B.C.), in *Riddles in Mathematics: A Book of Paradoxes:*

Protagoras made an arrangement with one of his pupils whereby the pupil was to pay for his instruction after he had won his first case. The young man completed his course, hung up the traditional shingle, and waited for clients. None appeared. Protagoras grew impatient and decided to sue his former pupil for the amount owed him.

''For,'' argued Protagoras, ''either I win this suit, or you win it. If I win, you pay me according to the judgment of the court. If you win, you pay me according to our agreement. In either case I am bound to be paid.''

''Not so,'' replied the young man. ''If I win, then by the judgment of the court I need not pay you. If you win, then by our agreement I need not pay you. In either case I am bound not have to pay you.''

The paradoxical statement is in fact one of the most powerful forms of language. It extends from the pun and Irish bull to the dramatic urgency of an oxymoron.

Sir Boyle Roche, a Dublin politician (1743–1807), was well known for Irish bulls:

''Half of the lies our opponents tell about us are not true.''
''A man cannot be in two places at once unless he is a bird.''
''I marvel at the strength of human weakness.''

Samuel Goldwyn, the famous motion picture magnate, used accidental paradoxes so often that Goldwynner became synonymous with the Irish bull:

''Include me out!''
''Didn't you hear me keeping still?''
''Our comedies are not to be laughed at.''

William Shakespeare often employed the oxymoron:

> Here's much to do with hate, but more with love.
> Why then, *O brawling love! O loving hate!*
> *O anything of nothing* first create!
> *O heavy lightness, serious vanity!*
> *Misshapen* chaos of *well-seeming* forms!
> *Feather of lead, bright smoke, cold fire, sick health!*
> *Still-waking sleep,* that is not what it is!
> This love I feel, that feel no love in this.

> ["Romeo and Juliet," I, i]

William Van O'Connor, "Paradox," in *Encyclopedia of Poetry and Poetics,* pointed out that the classical rhetoricians (e.g., Menander, Hermogenes, Cicero, and Quintilian) included paradox among the standard figures of speech. It was also popular during the Graeco-Roman period in such forms as paradoxical economium, *controversiae,* and *suasoriae,* was used to train students in rhetoric during the Middle Ages, and appears in the literature of the late Middle Ages and Renaissance (e.g., *The Praise of Folly* by Erasmus). During the baroque period, paradox became an important poetic figure (e.g., John Donne, *Paradoxes and Problems*). John Dryden and Alexander Pope both employed paradox, and William Hazlitt referred to Neoclassical poetry as "the poetry of paradox." During the twentieth century the term began to be applied to literary criticism (e.g., Cleanth Brooks, *The Well Wrought Urn*).

In a denotative sense, it is impossible for someone to be "awfully kind" or "terribly good" or for a relative to be "horribly decent." Of course, that is exactly what a poet wants, because it forces whoever reads or hears such statements to find a more subtle (connotative) connection, a "higher" truth.

But mathematical logic does not seek connotative truth; it seeks denotative explanations. Recreational mathematicians can enjoy the "human" side of, say, "The Barber's Paradox," chuckling over the contradiction the barber has gotten himself into, but serious mathematicians are troubled by it.

By the beginning of the twentieth century three major groups with different approaches to the solution of these paradoxes gathered around their leading figures: (1) the logistic group, led by Bertrand Russell; (2) the axiomatic group, led by David Hilbert; and (3) the intuitionist group, led by L. E. J. Brouwer.

The thrust of the logistic group has been to equate mathematics with symbolic logic. As can be seen from the above paradoxes, a circle occurs when a statement is made about *all* the members of a certain class of which the statement itself is a member. Bertrand Russell attempted to solve this problem by what he called the "theory of types." He put forth that logical statements, rules, things, and so on, are not all of the same type. Rather, they fit into a hierarchy of types that are very different from each other. Whatever category includes *all* of a class of things is not the same type of category as the things themselves. For example, the statement "All rules have exceptions" is not the same as the statement itself, which is a statement about statements about rules, or a rule about rules (rather than a rule about things).

The thrust of the axiomatic group is to base all of mathematics on a fundamental system of axioms (assumptions). The major problem for this group has been to prove that the axioms of arithmetic are consistent.

The thrust of the intuitionist group is that no mathematical concept is admissible unless it can be constructed. In other words, a mathematical concept must exist in more than name. In order for this to be true, a finite number of steps must be involved (e.g., all real numbers whose transfinite number is greater than a finite number is inadmissible).

None of these groups has been able to work its way free of paradoxes and other problems, and the work of Kurt Gödel in the 1930s seems to show that a consistent mathematics derived from a set of postulates through purely formal reasoning is impossible. Even within the most uncompromising formalistic, logical system, some form of constructive *intuition* remains.

Nevertheless, it is hard not to believe in Albert Einstein's famous statement: "God does not play at dice with the universe." Since we are unable to find an *absolute* proof for Einstein's statement, however, and especially in approaching the problem from the point of view of recreational mathematics, perhaps it is only fair to allow not Einstein but Lewis Carroll the final word. The following short piece comes from his miscellaneous writings, which have been collected in *The Complete Works of Lewis Carroll* (pp. 1230–31):

Which is better, a clock that is right only once a year, or a clock that is right twice every day? "The latter," you reply, "unquestionably." Very good, now attend.

I have two clocks: one doesn't go *at all,* and the other loses a minute a day: which would you prefer? "The losing," you answer, "without a doubt." Now observe: the one which loses a minute a day has to lose twelve hours, or seven hundred and twenty minutes before it is right again, consequently it is only right once in two years, whereas the other is evidently right as often as the time it points to comes round, which happens twice a day.

So you've contradicted yourself *once.*

"Ah, but," you say, "what's the use of its being right twice a day, if I can't tell when the time comes?"

Why, suppose the clock points to eight o'clock, don't you see that the clock is right *at* eight o'clock? Consequently, when eight o'clock comes round your clock is right.

"Yes, I see *that,*" you reply.

Very good, then you've contradicted yourself *twice:* now get out of the difficulty as best you can, and don't contradict yourself again if you can help it.

You *might* go on to ask, "How am I to know when eight o'clock *does* come? My clock will not tell me." Be patient: you know that when eight o'clock comes your clock is right, very good; then your rule is this: keep your eye fixed on your clock, and *the very moment it is right* it will be eight o'clock. "But—," you say. There, that'll do; the more you argue the farther you get from the point, so it will be as well to stop.

For similar mathematical recreations refer to Fallacies; Logical Problems and Puzzles; Zeno's Paradoxes.

BIBLIOGRAPHY

Bakst, Aaron. *Mathematics: Its Magic and Mastery.* New York: D. Van Nostrand Co., 1941.

Bell, E. T. *The Development of Mathematics.* New York: McGraw-Hill, 1940.

Carroll, Lewis. "The Two Clocks." N.d. Reprinted in *The Complete Works of Lewis Carroll.* New York: Random House, n.d.

Corbett, Edward P. J. *Classical Rhetoric for the Modern Student.* 2d ed. New York: Oxford University Press, 1971.

Courant, Richard, and Herbert Robbins. *What Is Mathematics?: An Elementary Approach to Ideas and Methods.* New York: Oxford University Press, 1958.

Gardner, Martin. *Aha! Gotcha: Paradoxes to Puzzle and Delight.* San Francisco: W. H. Freeman and Co., 1982.

————. *Martin Gardner's Sixth Book of Mathematical Games from "Scientific American."* San Francisco: W. H. Freeman and Co., 1971.

————. *Mathematical Carnival.* New York: Alfred A. Knopf, 1965.

Northrop, Eugene P. *Riddles in Mathematics: A Book of Paradoxes.* New York: D. Van Nostrand Co., 1944.

O'Connor, William Van. "Paradox." In *Encyclopedia of Poetry and Poetics.* Ed. by Alex Preminger. Princeton, N.J.: Princeton University Press, 1965.

Preminger, Alex, and Frank J. Warnke. "Oxymoron." In *Encyclopedia of Poetry and Poetics.* Ed. by Alex Preminger. Princeton, N.J.: Princeton University Press, 1965.

Russell, Bertrand. *Introduction to Mathematical Philosophy.* 2d ed. New York: Macmillan and Co., 1920.

————. and Alfred N. Whitehead. *Principia Mathematica.* 2d ed. Cambridge, Eng.: Cambridge University Press, 1935.

Schaaf, W. L. "Number Games and Other Mathematical Recreations." In *Macropaedia: Encyclopaedia Britannica,* 1985.

Shakespeare, William. "Romeo and Juliet." In *The Complete Signet Classic Shakespeare.* Ed. by Sylvan Bennet. New York: Harcourt Brace Jovanovich, 1963.

Shipley, Joseph T. *Playing with Words.* Englewood Cliffs, N.J.: Prentice-Hall, 1960.

**LOGICAL PROBLEMS AND PUZZLES** are those enigmas that do not involve the standard categories of mathematics (e.g., algebra) but nevertheless require the same form of reasoning as mathematics.

The unity of logic and pure mathematics was discovered in the nineteenth century. Until then, the various disciplines of mathematics were closely tied to the world as revealed through empiricism. Geometry, for instance, was viewed exclusively as a science of physical space, and the Euclidean axioms were considered as self-evidently true, so much so that Immanuel Kant and his predecessors maintained the existence of an *a priori* knowledge of nature.

In the early part of the nineteenth century Lobatchewsky was able to develop a (non-Euclidean) geometry that was as logically correct as Euclidean geometry, thus showing that the axioms of traditional Euclidean geometry were not *a priori* truths but, rather, merely convenient assumptions, which were neither true nor false in relation to physical space. In other words, pure geometry, as opposed to applied geometry, is incapable of deciding the question of whether physical space is Euclidean or not.

Rather, pure geometry is a branch of pure mathematics which serves only to develop the necessary consequences of various hypotheses or assumptions. Pure mathematics, then, does not deal with actual space but with classes of undefined terms and the relations between them. Its constraints and premises are none other than the formal constraints of logic. What is generally referred to as formal logic is nothing more than the study of the most general area of pure mathematics. (For more discussion of the logic of mathematics *see* Logical Paradoxes.)

Logical problems in recreational mathematics are generally solved by a form of trial and error. Examples of some of the more common forms follow.

One of the oldest forms of logical problem is that of *ferrying problems* or *difficult crossings*. W. L. Schaaf, ''Number Games and Other Mathematical Recreations,'' includes the following problem from Alcuin (c. 732–804): A husband and a wife and two children wish to cross a river in a boat that can hold only the husband or only the wife or only the children. How is this possible?

The solution follows: First the two children cross in the boat. Then one of the children returns in the boat. Then the husband crosses in the boat. Then the other child returns in the boat. Then the two children cross in the boat. Then one of the children brings the boat back. Then the wife crosses in the boat. Then one of the children brings the boat back. Then both the children cross in the boat.

Alcuin also offers the following problem: A man wishes to cross a river with a wolf, a goat, and a bundle of cabbages. However, the boat is so small that the man can take only one of them at a time (the wolf cannot be left alone with the goat; the goat cannot be left alone with the cabbages). How can this be accomplished?

Another age-old ferrying problem is generally called ''The Jealous Husbands.'' Three couples want to cross a river in a boat that will hold only two people. A woman is never to be left in the company of a man, unless her husband is present. How can this be accomplished?

Logical problems also include *truth and lies problems*. Here is an example: Some of the inhabitants of a town are pure-blooded English; the rest are half-English and half-French. The pure-blooded English always tell the truth; the half-English–half-French always lie. A visitor to the town is told that only one of three inhabitants he encounters is half-English–half-French. The visitor asks the three inhabitants whether they are pure-blooded English or half-English–half-French. The first inhabitant cannot be understood. The second points to the first and says, "He says that he is pure-blooded." The third inhabitant points to the second and says, "He lies." How is the visitor able to figure out which is half-and-half?

Here is another example. A mother returns from the store to find that one of her four children has broken a lamp in the livingroom. When she questions them, they respond as follows:

*Meghan:* Jay did it.

*Jay:* Ryan did it.

*Angie:* I didn't do it.

*Ryan:* Jay lied when he said I did it.

If only one of these statements is false, who was guilty? If only one is true, who was guilty?

Here is an example of what might be called a logical reversal problem, or turning the tables:

A father does not want a certain young man, Harry, to marry his daughter, Betty. However, Harry and Betty refuse to stay away from each other. In order to bring the matter to a conclusion, the father makes a proposition. "Children," he says, "I will make a deal with you. I will place two cards in my hat, a joker and a queen. Harry may then draw one of the cards from the hat. If he draws the joker, then he must leave and never return. If he draws the queen, then the two of you may marry."

"But, Father," Betty says, "that's not fair! What a dumb way to decide such a serious matter!"

"On the contrary," the father says, "it is the only fair way. Neither you nor I will give in to the other on this, so a completely impartial manner must be used to decide the matter. Now, the two of you go out on the back porch while I place the cards in the hat."

"But, Mother," Betty says, "Can't you see how unfair this is?"

"No!" the mother says. "I agree with your father. The two of you have continually talked your way around his wishes. You've used trickery time and again. It's time to have an end to the matter. There will be no more tricks. You will do as your father asks! Now go out on the porch and let your father prepare the hat."

Reluctantly, Harry and Betty go out on the porch. The father places two jokers in the hat and leaves the queen out of the hat. The mother does not know what the father has done, but Harry and Betty watch through the window. They return and begin to protest.

The mother cuts in, "No more! One more word and it's all over!"

"But, but," Harry says.

The father cuts him off, "Now, then, you must either accept my solution and draw a card or never see Betty again. If you decline or attempt to talk your way out of this, then I'll demand you never see each other again. *One more word out of either of you and Harry must leave!*"

What can be done? It seems hopeless. But Harry comes up with a solution. He reaches into the hat and pulls out a card and stuffs it in his mouth. The taste is terrible but he quickly chews and swallows it.

The only thing left is to look at the remaining card and deduce that the card chosen was the other card.

Joseph S. Madachy includes the same problem under a slightly different story line as "The Commoner's Dilemma," in *Mathematics on Vacation.*

Another form of logical problem involves overlapping sets. Here is an example: The college mathematics club included 14 students studying geometry, 11 studying algebra, 17 studying trigonometry, 5 studying both geometry and algebra, 7 studying both geometry and trigonometry, 9 studying both algebra and trigonometry, and 3 studying all three. How many members are there in the club, and how many are studying exactly two areas?

Here is one final classic (W. L. Schaaf includes it in his article). Three railway employees named Smith, Clark, and Jones work as brakeman, fireman, and engineer (though not necessarily in that order). They all live in the state of New York. Three businessmen with the same names also live in New York State. The businessman Clark and the brakeman live in Albany, the businessman Jones and the fireman live in Rochester, while the businessman Smith and the engineer live halfway between these two cities. The brakeman's namesake earns $3,500 a year; the engineer earns one-third as much as the businessman living nearest him. The railway man Smith beats the fireman at billiards. What is the engineer's name?

For similar discussions refer to Fallacies; Logical Paradoxes; Paradoxes of the Infinite; Zeno's Paradoxes.

BIBLIOGRAPHY

Bakst, Aaron. *Mathematics: Its Magic and Mastery.* New York: D. Van Nostrand Co., 1941.

Bell, E. T. *The Development of Mathematics.* New York: McGraw-Hill, 1940.

Cohen, Morris R. "Logic." In *The American People's Encyclopedia.* Ed. by Walter Dill Scott, Franklin J. Meine, and W. Stewart Wallace. Chicago: Spencer Press, 1953.

Courant, Richard, and Herbert Robbins. *What Is Mathematics? An Elementary Approach to Ideas and Methods.* New York: Oxford University Press, 1941.

Gardner, Martin. *Mathematical Carnival.* New York: Alfred A. Knopf, 1975.

Kline, Morris. *Mathematics in Western Culture.* New York: Oxford University Press, 1941.

Madachy, Joseph S. *Mathematics on Vacation.* New York: Charles Scribner's Sons, 1966.

Schaaf, W. L. ''Number Games and Other Mathematical Recreations.'' In *Macropaedia: Encyclopaedia Britannica,* 1985.

**LUCAS SEQUENCE.** See NUMBER PATTERNS, TRICKS, AND CURIOSITIES.

**LUCKY BONER.** See ILLEGAL OPERATION.

# M

**MAGIC CARD SQUARE.** See MAGIC SQUARES.

**MAGIC CIRCLES.** See MAGIC SQUARES.

**MAGIC CUBES.** See MAGIC SQUARES.

**MAGIC HEXAGON.** See MAGIC SQUARES.

**MAGIC SPHERES.** See MAGIC SQUARES.

**MAGIC SQUARES** consist of a number of integers arranged in the form of a square, so that the sum of each row, column, and main diagonal is equal.

If the integers used are the consecutive numbers from 1 to $n^2$ (as is the case with a standard magic square), then each magic square can be designated as a square of the $n$th order, and the sum of the numbers in each row, column, and main diagonal equals $\frac{1}{2}n(n^2 + 1)$.

It is thought that magic squares are of great antiquity, possibly originating in Babylonia or India and spreading to China. Helen Kunte, "Magic Squares and Circles," relates a traditional myth in China that when the emperor Yu (c. 2200 B.C.) stood by the Yellow River, there appeared a divine tortoise carrying two mystic symbols on its back, one of them a third order magic square.

A. Frederick Collins, *Fun with Figures,* claims that Archimedes (c. 250 B.C.) spent a great deal of time on magic squares, which, for this reason, have often been called the Square of Archimedes. C. A. Browne, "Magics and Pythagorean Numbers," in W. S. Andrews, ed., *Magic Squares and Cubes,* suggests that both Plato and Pythagoras knew about and used magic squares.

Paul Carus, "Reflections on Magic Squares," in Andrews, *Magic Squares and Cubes,* though suggesting that their use goes further back in history, finds the first indisputable magic squares appearing in China at the beginning of the Southern Sung dynasty (1127–1333). According to W. W. Rouse Ball, *Mathematical Recreations and Essays* and *A Short Account of the History of Mathematics,* Emmanuel Moschopulus, who lived in Constantinople in the early part of the fifteenth century, wrote a treatise on magic squares indicating that they can be definitely placed in Europe.

Cornelius Agrippa (1486–1535), one of the more famous of the astrologers who were impressed with magic squares, constructed magic squares of the orders 3, 4, 5, 6, 7, 8, and 9, which he associated in sequence with the seven heavenly bodies then known: Saturn, Jupiter, Mars, the Sun, Venus, Mercury, and the Moon. Furthermore, he asserted that a square of one cell represented the unity and eternity of God, and a square of two cells (which cannot be constructed) represented the imperfection of the four elements (air, earth, fire, and water); later writers would designate a square of two cells as a symbol of original sin.

A magic square engraved on a silver plate was often used as a charm against the plague. Albrecht Dürer's *Melancholy* (1514) includes a magic square, the numbers in the middle cells of the bottom row giving the date. Today, the magic square of the third order is worn throughout the East and India as a talisman.

There is only one possible arrangement of a third order magic square (it can, of course be rotated, but the numbers in relation to each other cannot be changed):

|   |   |   |
|---|---|---|
| 8 | 1 | 6 |
| 3 | 5 | 7 |
| 4 | 9 | 2 |

This is an *associated* or *regular* magic square, because the sum of any two numbers that are located in cells diametrically equidistant from the center of the square equals the sum of the first and last terms of the series, or $n^2 + 1$ (e.g., $3^2 + 1 = 10$, $8 + 2 = 10$, $1 + 9 = 10$, $6 + 4 = 10$, and $7 + 3 = 10$).

The next largest odd magic square is a fifth order or $5 \times 5$ square. There are numerous arrangements that satisfy this size magic square. The following is a common one where the sums of the rows, columns, and diagonals equal 65, and the sum of any two numbers equidistant from the center is 26, twice the center number:

| 17 | 24 | 1  | 8  | 15 |
|----|----|----|----|----|
| 23 | 5  | 7  | 14 | 16 |
| 4  | 6  | 13 | 20 | 22 |
| 10 | 12 | 19 | 21 | 3  |
| 11 | 18 | 25 | 2  | 9  |

One manner of constructing odd order magic squares is to begin with the first number in the center of the top row. Imagine the square circling back on itself and follow the diagonals up and to the right. Thus, 2 would be above the 8. Since the square circles back on itself, 2 ends up in the bottom row one column over from 1. Number 3, then, follows diagonally up to the right from 2; 4 follows up to the right from 3 (in this case, the square circles back on itself vertically); 5 follows 4 up and to the right. However, 1 is already where 6 would naturally occur. In that case, the next number drops straight down one row from the previous number. When a number reaches the upper right-hand corner, the next number would appear in the lower left-hand corner, unless a lower number is already there.

S. De la Loubère, *Du Royaume de Siam* (London, 1693, vol. ii, pp. 227–247), learned the above method while in Siam as an envoy of Louis XIV.

Eulerian squares are created by superposing two Latin squares of the $n$th order in such a way that the $n^2$ numbers (marks), each made up of two digits, will be different.

A Latin square is a square of $n^2$ cells (in $n$ rows and $n$ columns) in which $n^2$ letters (numbers) consisting of $n$ "a's," $n$ "b's," . . . are arranged so that $n$ letters in each row and each column are different:

|   |   |   |
|---|---|---|
| 2 | 0 | 1 |
| 1 | 2 | 0 |
| 0 | 1 | 2 |

In a diagonal Latin square, the letters (numbers) in the diagonals as well as the rows and columns are different.

By superposing the following two Latin squares on each other a Eulerian square of order three is formed:

| 2 | 0 | 1 |
|---|---|---|
| 1 | 2 | 0 |
| 0 | 1 | 2 |

| 2 | 1 | 0 |
|---|---|---|
| 1 | 0 | 2 |
| 0 | 2 | 1 |

| 22 | 01 | 10 |
|----|----|----|
| 11 | 20 | 02 |
| 00 | 12 | 21 |

The following diagonal Eulerian square solves the problem of the magic card square:

| 12 | 03 | 30 | 21 |
|----|----|----|----|
| 31 | 20 | 13 | 02 |
| 23 | 32 | 01 | 10 |
| 00 | 11 | 22 | 33 |

The magic card square problem is as follows: Take sixteen cards from a deck and arrange them in the form of a square so that no row, no column, and neither of the diagonals contains more than one card of each suit and one card of each rank.

There is an Eulerian square $(12, 11)_n$ for every odd value of $n$. By reversing the order of the rows and cyclically permuting the columns it is possible to derive the Eulerian square that underlies De la Loubère's rule. Thus, the De la Loubère magic square is a form of Eulerian square:

| 31 | 43 | 00 | 12 | 24 |
|----|----|----|----|----|
| 42 | 04 | 11 | 23 | 30 |
| 03 | 10 | 22 | 34 | 41 |
| 14 | 21 | 33 | 40 | 02 |
| 20 | 32 | 44 | 01 | 13 |

De la Loubère's method can be varied by placing number 1 in the cell horizontally to the right of the center cell and proceeding as before until a block is reached. In this case, the next number is placed in the second cell horizontally to the right of the last cell filled and the upward diagonal movement resumed:

| 4  | 29 | 12 | 37 | 20 | 45 | 28 |
|----|----|----|----|----|----|----|
| 35 | 11 | 36 | 19 | 44 | 27 | 3  |
| 10 | 42 | 18 | 43 | 26 | 2  | 34 |
| 41 | 17 | 49 | 25 | 1  | 33 | 9  |
| 16 | 48 | 24 | 7  | 32 | 8  | 40 |
| 47 | 23 | 6  | 31 | 14 | 39 | 15 |
| 22 | 5  | 30 | 13 | 38 | 21 | 46 |

It is also possible to form magic squares by employing the move of the chess piece, the knight, once again allowing the board to take on a circular property, as in the following (*see* Chess and Chess Problems for more discussion of the knight's move and knight's tours):

| 10 | 18 | 1  | 14 | 22 |
|----|----|----|----|----|
| 11 | 24 | 7  | 20 | 3  |
| 17 | 5  | 13 | 21 | 9  |
| 23 | 6  | 19 | 2  | 15 |
| 4  | 12 | 25 | 8  | 16 |

Starting in the middle of the top row, the knight moves up two spaces and over one, once again moving as if the top and bottom edges of the board were connected. If a move is blocked, as in the move after 5, the next number is placed in the space directly below the previous space.

It is also possible to build a magic square by moving four spaces to the right and one up:

| 80 | 58 | 45 | 23 | 1  | 69 | 47 | 34 | 12 |
|----|----|----|----|----|----|----|----|----|
| 9  | 68 | 46 | 35 | 11 | 79 | 57 | 44 | 22 |
| 10 | 78 | 56 | 43 | 21 | 8  | 67 | 54 | 32 |
| 20 | 7  | 66 | 53 | 31 | 18 | 77 | 55 | 42 |
| 30 | 17 | 76 | 63 | 41 | 19 | 6  | 65 | 52 |
| 40 | 27 | 5  | 64 | 51 | 29 | 16 | 75 | 62 |
| 50 | 28 | 15 | 74 | 61 | 39 | 26 | 4  | 72 |
| 60 | 38 | 25 | 3  | 71 | 49 | 36 | 14 | 73 |
| 70 | 48 | 35 | 13 | 81 | 59 | 37 | 24 | 2  |

And it is also possible to build one by moving two squares to the right and two down:

| 39 | 34 | 20 | 15 | 1  | 77 | 72 | 58 | 53 |
|----|----|----|----|----|----|----|----|----|
| 49 | 44 | 30 | 25 | 11 | 6  | 73 | 68 | 63 |
| 59 | 54 | 40 | 35 | 21 | 16 | 2  | 78 | 64 |
| 69 | 55 | 50 | 45 | 31 | 26 | 12 | 7  | 74 |
| 79 | 65 | 60 | 46 | 41 | 36 | 22 | 17 | 3  |
| 8  | 75 | 70 | 56 | 51 | 37 | 32 | 27 | 13 |
| 18 | 4  | 80 | 66 | 61 | 47 | 42 | 28 | 23 |
| 19 | 14 | 9  | 76 | 71 | 57 | 52 | 38 | 33 |
| 29 | 24 | 10 | 5  | 81 | 67 | 62 | 48 | 43 |

Magic squares of an even order are more difficult to build. Ball, *Mathematical Recreations and Essays,* divides even magic squares into two categories, those of a singly-even order ($n = 2(2m + 1)$), and those of a doubly-even order ($4m$). He credits Ralph Strachey for the following means of constructing those of a singly-even order:

First, divide the square into four quarters.

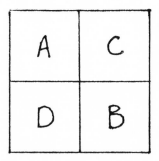

Second, construct a magic square in $\underline{A}$ by De la Loubère's method (possible because each of the four quarters of a singly-even order square will be an odd order square) with numbers 1 to $x^2$ where $x = n/2$. In a six order magic square, $\underline{A}$ would be a standard three order magic square.

Using the same rule, construct in <u>B</u>, <u>C</u>, and <u>D</u> similar magic squares $x^2 + 1$ to $2x^2$, $2x^2 + 1$ to $3x^2$, and $3x^2 + 1$ to $4x^2$.

| 8 | 1 | 6 | 26 | 19 | 24 |
|---|---|---|----|----|----|
| 3 | 5 | 7 | 21 | 23 | 25 |
| 4 | 9 | 2 | 22 | 27 | 20 |
| 35 | 28 | 33 | 17 | 10 | 15 |
| 30 | 32 | 34 | 12 | 14 | 16 |
| 31 | 36 | 29 | 13 | 18 | 11 |

In a six order square, $m$ equals 1, as can readily be obtained by plugging 1 into the formula stated above. The general procedure next requires the exchange of the $m$ squares next but one to the left-hand side of <u>A</u> (in this case, 5) with the same squares in <u>D</u> (in this case, 32) and the exchange of the $m$ squares nearest the left-hand side in all the other rows of <u>A</u> with their counterparts in <u>D</u> (in this case, 8 with 35 and 4 with 1). If $m$ were a higher number, say, 2, then the exchanges would be between the second and third squares in on the center row of <u>A</u> and the second and third squares in on the center row of <u>D</u>, and the first and second squares of <u>A</u> and <u>D</u> for the rest of the rows.

The final operation requires an exchange of the $m - 1$ (in this case, $m = 1$, so no exchange takes place) columns next to the right-hand side of <u>C</u> with the numbers in corresponding squares in <u>B</u>. The result is a magic square of the order six:

| 35 | 1 | 6 | 26 | 19 | 24 |
|----|---|---|----|----|----|
| 3 | 32 | 7 | 21 | 23 | 25 |
| 31 | 9 | 2 | 22 | 27 | 20 |
| 8 | 28 | 33 | 17 | 10 | 15 |
| 30 | 5 | 34 | 12 | 14 | 16 |
| 4 | 36 | 29 | 13 | 18 | 11 |

The following is an application of this operation to create a magic square of the order ten:

| 17 | 24 | 1 | 8 | 15 | 67 | 74 | 51 | 58 | 65 |
|----|----|----|----|----|----|----|----|----|----|
| 32 | 5 | 7 | 14 | 16 | 73 | 55 | 57 | 69 | 66 |
| 4 | 6 | 13 | 20 | 22 | 54 | 56 | 63 | 70 | 72 |
| 10 | 12 | 19 | 21 | 3 | 60 | 62 | 69 | 71 | 53 |
| 11 | 18 | 25 | 2 | 9 | 61 | 68 | 75 | 52 | 59 |
| 92 | 99 | 76 | 83 | 90 | 42 | 49 | 26 | 33 | 40 |
| 98 | 80 | 82 | 89 | 91 | 48 | 30 | 32 | 39 | 41 |
| 79 | 81 | 88 | 45 | 97 | 29 | 31 | 38 | 45 | 47 |
| 85 | 87 | 94 | 96 | 78 | 35 | 37 | 44 | 96 | 28 |
| 86 | 93 | 100 | 77 | 84 | 36 | 43 | 50 | 27 | 34 |

| 92 | 99 | 1 | 8 | 15 | 67 | 74 | 51 | 58 | 40 |
|----|----|----|----|----|----|----|----|----|----|
| 98 | 80 | 7 | 14 | 16 | 73 | 55 | 57 | 64 | 41 |
| 4 | 81 | 88 | 20 | 22 | 54 | 56 | 63 | 70 | 47 |
| 85 | 87 | 19 | 21 | 3 | 60 | 62 | 69 | 71 | 28 |
| 86 | 93 | 25 | 2 | 9 | 61 | 68 | 75 | 52 | 34 |
| 17 | 24 | 76 | 83 | 90 | 42 | 49 | 26 | 33 | 65 |
| 32 | 5 | 82 | 89 | 91 | 48 | 30 | 32 | 39 | 66 |
| 79 | 6 | 13 | 95 | 97 | 29 | 31 | 38 | 45 | 72 |
| 10 | 12 | 94 | 96 | 78 | 35 | 37 | 44 | 46 | 53 |
| 11 | 18 | 100 | 77 | 84 | 36 | 43 | 50 | 27 | 59 |

A magic square of a doubly-even order (4m) can be constructed by writing the numbers sequentially from the upper left-hand corner across and replacing the numbers on the diagonal squares of every component block of sixteen squares by their complements. In a fourth order square there is only one set of diagonals:

By replacing the numbers on these diagonals with the numbers in complementary squares, a magic square of the fourth order is obtained:

| 16 | 2  | 3  | 13 |
|----|----|----|----|
| 5  | 11 | 10 | 8  |
| 9  | 7  | 6  | 12 |
| 4  | 14 | 15 | 1  |

If an order eight square is desired, the same process is employed:

|    | 2  | 3  |    |    |    | 6  | 7  |    |
|----|----|----|----|----|----|----|----|
| 9  |    |    | 12 | 13 |    |    | 16 |
| 17 |    |    | 20 | 21 |    |    | 24 |
|    | 26 | 27 |    |    | 30 | 31 |    |
|    | 34 | 35 |    |    | 38 | 39 |    |
| 41 |    |    | 44 | 45 |    |    | 48 |
| 49 |    |    | 52 | 53 |    |    | 56 |
|    | 58 | 59 |    |    | 62 | 63 |    |

| 64 | 2  | 3  | 61 | 60 | 6  | 7  | 57 |
|----|----|----|----|----|----|----|----|
| 9  | 55 | 54 | 12 | 13 | 51 | 50 | 16 |
| 17 | 47 | 46 | 20 | 21 | 43 | 42 | 24 |
| 40 | 26 | 27 | 37 | 36 | 30 | 31 | 33 |
| 32 | 34 | 35 | 29 | 28 | 38 | 39 | 25 |
| 41 | 23 | 22 | 44 | 45 | 19 | 18 | 48 |
| 49 | 15 | 14 | 52 | 53 | 11 | 10 | 56 |
| 8  | 58 | 59 | 5  | 4  | 62 | 63 | 1  |

Benjamin Franklin produced an 8 × 8 magic square (see the following illustration) and a 16 × 16 magic square with additional interesting properties. The magic constant for the 8 × 8 square is 260. Every half-row and half-column equals 130. The four corners plus the middle four squares total 260. The bent row of eight numbers ascending and descending diagonally, e.g., from 16 ascending to 10, and from 23 descending to 17, and every one of its parallel bent rows of eight numbers, equals 260. The bent row from 52 descending to 54, and from 43 ascending to 45, and every one of its parallel bent rows of eight numbers, makes 260. The bent row from 45 to 43, descending to the left, and from 23 to 17, descending to the right, and each of its parallel bent rows of eight numbers, equals 260. The bent row from 52 to 54, descending to the right, and from 10 to 16, descending to the left, and each of its parallel bent rows of eight numbers, makes 260. The parallel bent rows next to those previously mentioned, shortened to three numbers ascending and three descending, make, with the two corner numbers, 260 (e.g., from 53 to 4 ascending and from 29 to 44 descending). The two numbers, 14 and 61, ascending, and 36 and 19, descending, with the four numbers similar to them, e.g., 50 and 1, descending, and 32 and 47, ascending, makes 260.

| 52 | 61 | 4 | 13 | 20 | 29 | 36 | 45 |
|----|----|----|----|----|----|----|----|
| 14 | 3 | 62 | 51 | 46 | 35 | 30 | 19 |
| 53 | 60 | 5 | 12 | 21 | 28 | 37 | 44 |
| 11 | 6 | 59 | 54 | 43 | 38 | 27 | 22 |
| 55 | 58 | 7 | 10 | 23 | 26 | 39 | 42 |
| 9 | 8 | 57 | 56 | 41 | 40 | 25 | 24 |
| 50 | 63 | 2 | 15 | 18 | 31 | 34 | 47 |
| 16 | 1 | 64 | 49 | 48 | 33 | 32 | 17 |

Joseph S. Madachy, *Mathematics on Vacation,* also claims that Franklin's square is diabolic (i.e., the broken diagonals also total the magic number). This just is not the case, however (e.g., the broken diagonal 53-3-4-49-18-40-39-22 = 228).

Diabolic magic squares (also referred to as panmagic or Nasik squares) were first discovered by Rev. A. W. Frost and named Nasik after the town in India where he resided. Here is an example:

| 1 | 14 | 7 | 12 |
| 15 | 4 | 9 | 6 |
| 10 | 5 | 16 | 3 |
| 8 | 11 | 2 | 13 |

In this example, the rows, the columns, and the diagonals, including the broken diagonals (e.g., 10-4-7-13) equal 34.

A bordered magic square, as the name implies, has the additional property of remaining a magic square when the outside rows are removed from each side. Here is an example:

| 4 | 5 | 6 | 43 | 39 | 38 | 40 |
| 49 | 15 | 16 | 33 | 30 | 31 | 1 |
| 48 | 37 | 22 | 27 | 26 | 13 | 2 |
| 47 | 36 | 29 | 25 | 21 | 14 | 3 |
| 8 | 18 | 24 | 23 | 28 | 32 | 42 |
| 9 | 19 | 34 | 17 | 20 | 35 | 41 |
| 10 | 45 | 44 | 7 | 11 | 12 | 46 |

Magic squares have also been constructed using only prime numbers. Ball, *Mathematical Recreations and Essays,* includes one of the order three from H. E. Dudeney and one of the order four from E. Bergholt and C. D. Shuldham:

| 67 | 1 | 43 |
| 13 | 37 | 61 |
| 31 | 73 | 7 |

| 3 | 71 | 5 | 23 |
| 53 | 11 | 37 | 1 |
| 17 | 13 | 41 | 31 |
| 29 | 7 | 19 | 47 |

It is also possible to construct magic squares so that when the number of each square is replaced by its square the magic square will still be magic. Here is an example of a magic square that is also a diabolic square and remains a magic square if each number in it is squared. It comes from Ball, who attributes it to M. H. Schots:

| 17 | 50 | 43 | 04 | 32 | 75 | 66 | 21 |
|----|----|----|----|----|----|----|----|
| 31 | 76 | 65 | 22 | 14 | 53 | 40 | 07 |
| 00 | 47 | 54 | 13 | 25 | 62 | 71 | 36 |
| 26 | 61 | 72 | 35 | 03 | 44 | 57 | 10 |
| 45 | 02 | 11 | 56 | 60 | 27 | 34 | 73 |
| 63 | 24 | 37 | 70 | 46 | 01 | 12 | 55 |
| 52 | 15 | 06 | 41 | 77 | 30 | 23 | 64 |
| 74 | 33 | 20 | 67 | 51 | 16 | 05 | 42 |

It is also possible to create magic squares which are magic for the original numbers, the numbers squared, and the numbers cubed. The smallest squares so far formed of this type are of the order 64 (by E. Cazalas, and by Royal V. Heath).

Magic cubes of the $n$th order consist of consecutive numbers from 1 to $n^3$ arranged in the shape of a cube, so that the sum of the numbers in every row, every column, every file, and each of the four diagonals is equal, i.e., $\frac{1}{2}n(n^3 + 1)$.

The smallest magic cube is $3 \times 3 \times 3$. The following example is found in "Magic Cubes," Andrews, *Magic Squares and Cubes*. The nine different squares it contains are placed in columns:

| Column I | Column II | Column III |
|----------|-----------|------------|
| Top to Bottom | Front to Back | Left to Right |

| 10 | 26 | 6 |
|----|----|----|
| 24 | 1 | 17 |
| 8 | 15 | 19 |

| 8 | 15 | 19 |
|----|----|----|
| 12 | 25 | 5 |
| 22 | 2 | 18 |

| 10 | 24 | 8 |
|----|----|----|
| 23 | 7 | 12 |
| 9 | 11 | 22 |

| 23 | 3 | 16 |
|----|----|----|
| 7 | 14 | 21 |
| 12 | 25 | 5 |

| 24 | 1 | 17 |
|----|----|----|
| 7 | 14 | 21 |
| 11 | 27 | 4 |

| 26 | 1 | 15 |
|----|----|----|
| 3 | 14 | 25 |
| 13 | 27 | 2 |

| Column I | Column II | Column III |
|---|---|---|
| Top to Bottom | Front to Back | Left to Right |

| 9 | 13 | 20 |
|---|---|---|
| 11 | 27 | 4 |
| 22 | 2 | 18 |

| 10 | 26 | 6 |
|---|---|---|
| 23 | 3 | 16 |
| 9 | 13 | 20 |

| 6 | 17 | 19 |
|---|---|---|
| 16 | 21 | 5 |
| 20 | 4 | 18 |

In this cube there are twenty-seven straight columns, two diagonal columns in each of the three middle squares, and four diagonal columns connecting the eight corners of the cube, equalling thirty-seven columns, each of which adds up to 42. The center number is 14 $(n^3 + 1)/2$, and the sum of diametrically opposite numbers is 28 $(n^3 + 1)$.

This cube is built as follows: First, "breakmoves," or moves which do not proceed according to the normal pattern, occur between each multiple of $n^2$ and the next number (i.e., between 9 and 10, and between 18 and 19, and between 27 and 1), and between all other multiples of $n$ (i.e., between 3 and 4, between 6 and 7, between 12 and 13, between 15 and 16, between 21 and 22, and between 24 and 25).

Begin by placing 1 in the center of the top square. Then move one cell down in the next square up (consider the cube as a three-dimensional cylinder, just as in the two-dimensional cylinder of a square, so that a move up from the top square is a move onto the bottom square). Thus, 2 will be in the bottom center cell of the bottom square. Continue in this manner until a breakmove for a multiple of $n^2$ is necessary (in which case move to the same cell in the next square down from the last entry) or a breakmove for a multiple of $n$ is reached (in which case move one cell in a downward right-hand diagonal in the next square down from the last entry).

The cube can also be constructed by using either the second or the third columns. If the second column is used, the rules are as follows: Begin with 1 in the middle cell of the upper row of numbers in the middle square. Each move is one cell up in the next square up. A breakmove for $n^2$ is one cell down in the same square. A breakmove for $n$ is one cell in a downward right-hand diagonal in the next square up.

If the third column is used, the rules are as follows: Begin with 1 in the middle cell of the upper row of numbers in the middle square. Each move is three consecutive cells in an upward right-hand diagonal in the same square. A breakmove for $n^2$ is one cell down in the same square. A breakmove for $n$ is one cell in a downward right-hand diagonal in the next square down.

A $5 \times 5 \times 5$ cube can be constructed by following these directions: Begin with 1 in the first cell of the middle horizontal column in the third square. Each move is one cell up in the next square down. For each breakmove for $n^2$ move one cell to the right in the same square as the last entry. For each breakmove for *n* move two cells to the left and one cell down (a knight's move) in the same square as the last entry.

Magic configurations go beyond magic squares and cubes to other plane and solid figures. Magic circles are circles whose intersecting points are numbered so that the total of the numbers for each circle is equal. The following example is the result of flattening the circular intersections of a standard die:

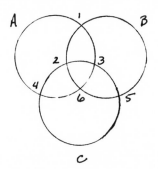

Circle *A* has numbers 1-3-6-4 at its points of intersection, a total of 14. Circle *B* has 1-2-6-5 at its points of intersection, a total of 14. Circle *C* has 4-2-3-5 at its points of intersection, a total of 14.

Much more complex circles are possible. Here is an example:

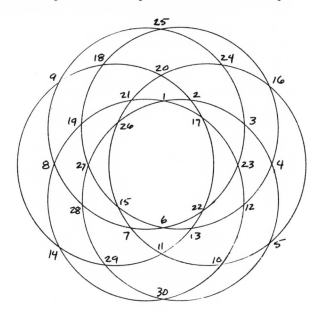

Magic spheres can be created by placing complementary numbers diametrically opposite, thus making them associated. Here is an example:

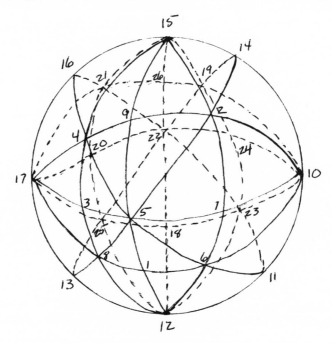

If the circles are not all equal "great" circles, other rules apply in addition to placing complementary numbers diametrically opposite.

Frederick A. Morton, "Magic Stars," *in Andrews, Magic Circles and Cubes,* offers the following rules for constructing magic stars of all orders.

Multiply a chosen summation (e.g., 48) by the order of star used (e.g., 5) and divide by 2 to obtain the summation of the series (in this case, $(5 \times 48)/2 = 120$).

Divide this into two parts (e.g., 80 and 40). Select one of the two parts, say, 40, and find a series of five numbers whose sum equals it, e.g., $6 + 7 + 8 + 9 + 10 = 40$. These numbers are then written in the central pentagon of the star as indicated:

Add the numbers sequentially around the pentagon:

$6 + 9 = 15$
$7 + 10 = 17$
$8 + 6 = 14$
$9 + 7 = 16$
$10 + 8 = 18$

List the totals in a separate column:

15
17
14
16
18

Place on each side of the first number (15) numbers not yet used in the central pentagon which will total the three numbers at 48 (the chosen summation): $17 + 15 + 16 = 48$.

Place the last number of those added (16) under the first (17), and under the first 16 place the number needed to once again reach a total of 48. Continue in this manner through all of the numbers:

$17 + 15 + 16 = 48$
$16 + 17 + 15 = 48$
$15 + 14 + 19 = 48$
$19 + 16 + 13 = 48$
$13 + 18 + 17 = 48$

If the correct numbers were selected for the first trio, then the first number of the first trio will match the final number of the final trio, and the star will sum correctly if the numbers in the first column are used to designate the points of the star, as shown:

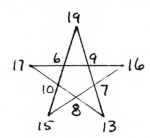

If the first and last numbers of the trios are different, they can be corrected as follows: If the final number is more than the first number, add half the difference between the two numbers to the first number and proceed as before. If the final number is less than the first number, subtract half the difference from the first number and proceed as before.

According to Joseph S. Madachy, there is only one possible magic hexagon (which, of course, can be rotated and reflected):

Helen Knute, "Magic Squares and Circles," includes examples of Japanese magic circles (they use a different set of rules than the magic circles previously discussed). $N$ concentric circles are divided into equal parts by $n$ diameters. The object, then, is to place consecutive integers at the intersections so that the sum of the digits along each circle will be the same; 1 is placed at the center of the circle:

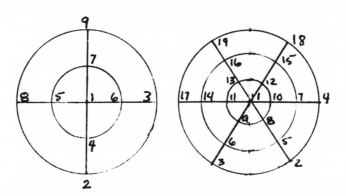

The constant sum in the first circle is 22, in the second 63.

Magic squares (and other configurations) can involve operations other than addition. Henry E. Dudeney was, perhaps, the first to explore these other possibilities:

A *subtracting magic square*:

| 2 | 1 | 4 |
|---|---|---|
| 3 | 5 | 7 |
| 6 | 9 | 8 |

By subtracting the center number in each row, column, and diagonal from the two outside numbers added together, 5 is always obtained.

As *multiplying magic square*:

| 12 | 1 | 18 |
|---|---|---|
| 9 | 6 | 4 |
| 2 | 36 | 3 |

By multiplying every row, column, and diagonal, 216 is always obtained.

A *dividing magic square*:

| 3 | 1 | 2 |
|---|---|---|
| 9 | 6 | 4 |
| 18 | 36 | 12 |

By dividing the second number in a line by the first in either direction and the third number by the quotient or the product of the two extremes by the middle number, 6 is always obtained.

Collins, *Fun with Figures,* includes the following magic square based on fractions:

| $\frac{3}{8}$ | $\frac{5}{12}$ | $\frac{5}{24}$ |
|---|---|---|
| $\frac{1}{6}$ | $\frac{1}{3}$ | $\frac{1}{2}$ |
| $\frac{11}{24}$ | $\frac{1}{4}$ | $\frac{7}{24}$ |

When the fractions are added together by rows, columns, or diagonals, they equal 1.

Antimagic squares are squares containing an array of integers from 1 to $n^2$ so that each row, column, and main diagonal produces different sums and these sums are part of a consecutive series of integers (they are a special form of heterosquare, a square of integers from 1 to $n^2$ such that all the rows, columns, and main diagonals have different sums).

Heterosquares of order two are impossible. Charles W. Trigg found the following heterosquare of order three:

|   |   |   |    |
|---|---|---|----|
| 9 | 8 | 7 | 24 |
| 2 | 1 | 6 | 9  |
| 3 | 4 | 5 | 12 |

11  14  13  18  15

One method of creating heterosquares of order four and up is to list the numbers consecutively in a square and then reverse the final two numbers:

|    |    |    |    |
|----|----|----|----|
| 1  | 2  | 3  | 4  |
| 5  | 6  | 7  | 8  |
| 9  | 10 | 11 | 12 |
| 13 | 14 | 16 | 15 |

It might be interesting to compare this phenomenon with theories of polarity (refer to Nim and The Fifteen Puzzle).

There is no systematic approach (to my knowledge) to the construction of antimagic squares. It is believed that antimagic squares of orders one, two, and three are impossible. Here is an example of a fourth order antimagic square:

|    |    |    |    |    |
|----|----|----|----|----|
| 10 | 14 | 5  | 1  | 30 |
| 12 | 2  | 13 | 6  | 33 |
| 3  | 15 | 9  | 11 | 38 |
| 7  | 8  | 4  | 16 | 35 |

36  32  39  31  34  37

For similar play refer to Chess; Chess Problems; Fairy Chess; The Fifteen Puzzle: Nim; Topology.

BIBLIOGRAPHY

Andrews, W. S., ed. *Magic Squares and Cubes.* 2d ed., Open Court Publishing Company, 1917. Reprint. New York: Dover, 1960.

Ball, W. W. Rouse. *Mathematical Recreations and Essays.* 1892. Reprint. London: Macmillan and Co., 1939.

————. *A Short Account of the History of Mathematics.* 1888. Reprint. London: Macmillan and Co., 1935.

Bowers, Henry, and Joan E. Bowers. *Arithmetical Excursions: An Enrichment of Elementary Mathematics.* New York: Dover, 1961.

Collins, A. Frederick. *Fun with Figures.* New York: D. Appleton and Co., 1930.

Dudeney, Henry Ernest. *Amusements in Mathematics.* 1917. Reprint. New York: Dover, 1958.

Friend, J. Newton. *More Numbers: Fun and Facts.* New York: Charles Scribner's Sons, 1961.

Gardner, Martin. *The Second "Scientific American" Book of Mathematical Puzzles and Diversions.* New York: Simon & Schuster, 1961.

Kraitchik, Maurice. *Mathematical Recreations.* 2d rev. ed. New York: Dover, 1953.

Kunte, Helen. "Magic Squares and Circles." In *The National Encyclopedia,* 1944.

Madachy, Joseph S. *Mathematics on Vacation.* New York: Charles Scribner's Sons, 1966.

Schaaf, W. L. "Number Games and Other Mathematical Recreations." In *Macropaedia: Encyclopaedia Britannica,* 1985.

Simon, W. *Mathematical Magic.* New York: Charles Scribner's Sons, 1964.

**MAGIC STARS.** See MAGIC SQUARES.

**MAKING THE RIGHT MISTAKE.** See ILLEGAL OPERATION.

**MANCALA.** See NIM.

**MAP COLORING PROBLEM.** See TOPOLOGY.

**MARKED-PAWN GAME.** See FAIRY CHESS.

**MARSEILLE CHESS.** See FAIRY CHESS.

**MARSEILLE GAME.** See FAIRY CHESS.

**MATADOR.** See DOMINOES.

**MATCH PROBLEMS, PUZZLES, AND GAMES** involve the manipulation of matches for mathematical recreation.

*Maxey* is a game played with matches. Seven lines are drawn on a piece of paper about the length of a match and a match length apart. Each player is given five matches and sits on opposite sides of the lines (the lines pointing at the players). Each player plays one match at a time, pointing the match head toward himself. If two matches are placed side by side, a player may place a match across the two (the match head pointing to the player's right). One point is scored by placing a match adjacent to a played match, and two points are scored for playing a match across two played matches. The player with the most points wins.

A. Frederick Collins includes the following two match tricks in *Fun with Figures*: In the first trick, three matches are placed on a table. The person doing the trick then picks up and returns the matches to the table, counting one, two, three as he does. The same is done for the fourth match. However, when the fifth match is picked up it is held in the hand. Then the sixth and seventh matches are picked up and counted. Then the three matches in the hand are laid back on the table and counted, eight, nine, ten. Nine matches picked up three times have equalled ten.

The second trick (or puzzle) is a play on words. Two triangles are formed with five matches, as follows:

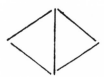

Then the unsuspecting game player, who has probably been puzzling through match tricks for some time, is told to remove three matches and replace them in such a manner that a pair of like triangles will be formed.

When the player admits defeat, the person playing the trick simply removes the three matches from the right and places them in the same position a short distance away. Then he takes the other two matches and places them against the first three exactly as before, satisfying the conditions of the problem and producing a groan from all onlookers.

Here are some more legitimate puzzles: In the first, the object is to make exactly four equilateral triangles using six identical matches. The solution is as follows:

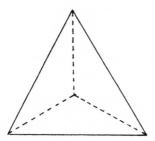

The key is to think in three dimensions. In the following the object is to move three of the matches to create three squares using all of the matches:

The solution is as follows:

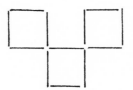

Two matches have been moved from the lower right-hand corner and one from the upper left top to form the third square to the left of the previous squares.

In the following the object is to move (not remove) three matches and leave only five squares:

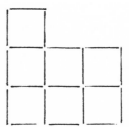

The solution is as follows:

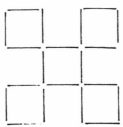

The following problems come from Boris A. Kordemsky, *The Moscow Puzzles:* In the first the object is to place 26 matches on the table to form three squares:

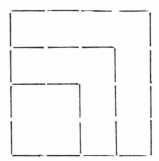

In the second the object is to form as many identical squares as possible with twenty-four matches. The possibilities are as follows:

1. By placing three matches to a side it is possible to form two squares.
2. By placing two matches to a side it is possible to form three squares.

3. By overlapping two squares with three matches to a side it is possible to form three squares (though they are not all the same size):

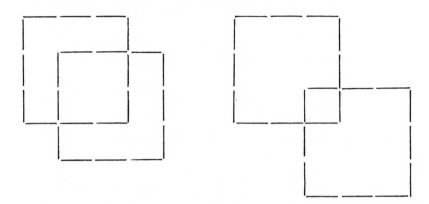

4. By overlapping as follows, seven squares are possible (not all of the same size):

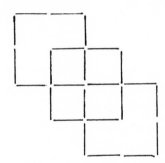

5. By using one match to a side, numerous patterns are possible, obtaining as many as fourteen squares (not all of the same size). Some of the patterns follow:

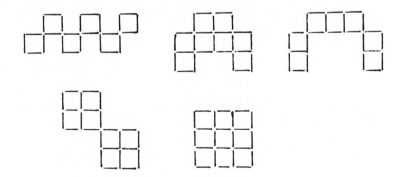

6. By laying matches on top of other matches, the following formations are possible:

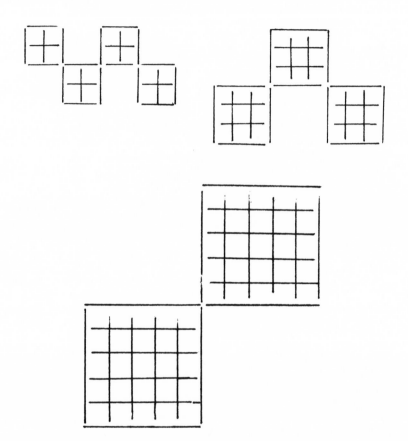

For similar play refer to Geometric Dissections; Geometric Problems and Puzzles; Nim.

BIBLIOGRAPHY

Brandreth, Gyles. *The World's Best Indoor Games*. New York: Pantheon Books, 1981.
Collins, A. Frederick. *Fun with Figures*. New York: D. Appleton and Co., 1928.
Dudeney, Henry Ernest. *536 Puzzles and Curious Problems*. Reprint. Ed. by Martin Gardner. New York: Dover, 1967.
Friend, J. Newton. *Still More Numbers: Fun and Facts*. New York: Charles Scribner's Sons, 1964.
Gardner, Martin. *Mathematical Circus*. New York: Alfred A. Knopf, 1979.

James, Glenn, ed. *The Tree of Mathematics*. California: The Digest Press, 1957.

Kordemsky, Boris A. *The Moscow Puzzles*. Trans. by Albert Parry, ed. by Martin Gardner. New York: Charles Scribner's Sons, 1972.

**MAXEY.** See MATCH PROBLEMS, PUZZLES, AND GAMES.


**MAXIMUMMER.** See FAIRY CHESS.


**MAZES AND LABYRINTHS** are topological puzzles that involve getting from one point through a number of passages to another point.

The term "maze" may have come from the verb "amazed" (lost in thought). "Labyrinth" comes from Greek for passages of a mine.

The most famous labyrinth is the one built by Daedalus in Crete for King Minos. The Minotaur, offspring of Minos' wife Pasiphae and a white bull with a black spot between its horns, was kept hidden at the labyrinth's center. Anyone who entered the labyrinth became lost in its passages and was devoured by the Minotaur. Seven youths and seven maidens were periodically sent from Athens to appease the monster, until Theseus, with the aid of a thread provided by Ariadne, entered the labyrinth, slew the Minotaur, and found his way back out by following the thread. According to Henry Ernest Dudeney, this is where the phrase "threading a maze" comes from.

Whether or not Daedalus' labyrinth in Crete ever really existed is unknown. No ruins have been found, but ancient writers place its existence near Knossos. Egyptologist Karl R. Lepsius found the remains of another ancient labyrinth in Egypt, about 1,000 feet long by 800 feet wide, which was probably built by order of Amenemhat III (2300 B.C.) and used as his tomb. A similar ancient labyrinth has been found at Lemnian, and another, the Italian Labyrinth, which served as the tomb of Porsena near Clusium, has been found near Chiusi.

By the twelfth century, Christian cathedrals and churches were using labyrinths in a symbolic sense (perhaps as a symbol of the maze of sin man had to find his way through to reach heaven), and a number of labyrinths were built around the cathedrals and churches as a literal expression of this symbol.

Examples of these include the parish church at St. Quentin, the abbey church of St. Bertin at St. Omer, Chartres Cathedral, Amiens Cathedral, Lucca Cathedral, the Abbey of Toussarts, St. Michele at Pavia, and the cathedrals of Poitiers and Rheims.

Twelfth-century labyrinths were also built for less holy reasons. England's most famous twelfth-century labyrinth was built by Rosamond Bower in a park at Woodstock for King Henry II, who sought to conceal his mistress, Rosamond the

Fair, from his wife, Eleanor of Aquitaine. However, using Ariadne's string technique, Eleanor found Rosamond the Fair and forced her to drink poison.

After the Reformation, labyrinths became a popular form of garden landscaping called "puzzle gardens." Examples are the Hampton Court Maze, the Hatfield House Maze, and the Pimperne Maze.

A labyrinth or maze is either "simply connected" (i.e., having no detached walls) or "multiply connected" (i.e., having detached walls). If the maze has only one entrance and one exit, it can be traversed (though not necessarily by the shortest path) by keeping one hand on one of the walls. This technique also works for locating a place within the maze, providing there is no "closed circuit" (route that allows a complete path back to the beginning).

A standard method for solving a maze (Dudeney gives a M. Trémaux credit for it) goes as follows: First, never pass over a path more than twice; second, select either path when arriving at a new node (place of new turning); third, when arriving at an old node or dead end, return by the same path; fourth, when arriving at an old node by an old path, select a new path, if possible.

Refer to Geometric Problems and Puzzles for similar recreations.

BIBLIOGRAPHY

Bright, Greg. *The Great Maze Book: Extraordinary Puzzles for Extraordinary People.* New York: Random House, 1973.

Dudeney, Henry Ernest. *Amusements in Mathematics.* 1917. Reprint. New York: Dover, 1958.

Grant, Michael. *Myths of the Greeks and Romans.* New York: New American Library, 1962.

Kinnard, Clark, ed. *Encyclopedia of Puzzles and Pastimes.* New York: Citadel Press, 1946.

Schaaf, W. L. "Number Games and Other Mathematical Recreations." In *Macropaedia: Encyclopaedia Britannica,* 1985.

**MERSENNE NUMBERS.** See NUMBER PATTERNS, TRICKS, AND CURIOSITIES; THE TOWER OF HANOI.

**MIRRORS** include mathematical encounters with inversion and reflection.

"Now, if you'll only attend, Kitty, and not talk so much, I'll tell you all my ideas about Looking-glass House. First, there's the room you can see through the glass—that's just the same as our drawing-room, only the things go the other way." [Lewis Carroll, *Through the Looking-Glass: And What Alice Found There,* p. 134]

*He took his vorpal sword in hand:*
*Long time the manxome foe he sought—*
*So rested he by the Tumtum tree,*
*And stood awhile in thought.*

Alice Raikes, Lewis Carroll's distant cousin, explains how the mirror idea got into *Through the Looking-Glass* ([London] *Times,* January 22, 1932; reprint in Martin Gardner, ed., *The Annotated Alice,* p. 180):

As Children, we lived in Onslow Square and used to play in the garden behind the houses. Charles Dodgson used to stay with an old uncle there, and walk up and down, his hands behind him, on the strip of lawn. One day, hearing my name, he called me to him saying, "So you are another Alice. I'm very fond of Alices. Would you like to come and see something which is rather puzzling?" We followed him into his house which opened, as ours did, upon the garden, into a room full of furniture with a tall mirror standing across one corner.

"Now," he said, giving me an orange, "first tell me which hand you have got that in." "The right," I said. "Now," he said, "go and stand before that glass, and tell me which hand the little girl you see there has got it in." After some perplexed contemplation, I said, "The left hand." "Exactly," he said, "and how do you explain that?" I couldn't explain it, but seeing that some solution was expected, I ventured, "If I was on the *other* side of the glass, wouldn't the orange still be in my right hand?" I can remember his laugh. "Well done, little Alice," he said. "The best answer I've had yet."

I heard no more then, but in after years was told that he said that had given him his first idea for *Through the Looking-Glass,* a copy of which, together with each of his other books, he regularly sent me.

As Martin Gardner points out, the strange properties of the looking-glass or mirror serve as a central theme in the book, in particular the theme of *inversion.* Here are a few of the examples he lists: Tweedledee and Tweedledum are mirror-image twins; the White Knight sings of forcing his right foot into a left shoe; Alice walks backward to approach the Red Queen; the looking-glass cake is handed around first, then sliced; the White Queen explains the advantages of living backward in time; and so on.

Lewis Carroll was a mathematician, so the use of inversion in his writings is not surprising. As Bonnie Averbach and Orin Chein explain in *Mathematics: Problem Solving Through Recreational Mathematics,* one form of mathematical inversion occurs anytime an ordering of numbers (a permutation) has a larger number preceding a smaller number, e.g., in the permutation of the numbers 1, 2, . . . $n$, an inversion occurs once in the sequence 1, 2, 3, 5, 4 and four times in the sequence 1, 5, 2, 4, 3 (5 precedes 2, 4, and 3; 4 precedes 3). This introduces the concept of polarity, which is central to such important mathematical games as Nim and to puzzles such as the Fifteen Puzzle (refer to these entries for more in-depth discussion of polarity).

Aaron Bakst, *Mathematics: Its Magic and Mastery,* points out another form of inversion in mathematics, the inverse operations of arithmetic: addition/ subtraction, multiplication/division, e.g., $2 + 2 = 4$, $4 - 2 = 2$; and $2 \times 3 = 6$, $6/3 = 2$. A special form of this occurs when numbers are multiplied by themselves, i.e., are squared. The inverse of this is called the extraction of square roots, e.g., $15^2 = 225$. The inverse of this is $\sqrt{225} = \sqrt{15^2} = 15$.

Inversion at this level is interesting and expresses a satisfying symmetry. However, when inversion occurs during the transformation or mapping of other planes, things get more complex. For example, consider the inversions that take place in respect to circles (sometimes called circular reflections because they represent the relation between an original shape and its image in a circular mirror). On a fixed plane, let $C$ be a given circle with center $O$ (the center of the inversion) and radius $r$. Then, the image of any point $P$ is defined as point $P'$ which lies on the line $OP$ on the same side of $O$ as $P$ so that $OP \cdot OP' = r^2$. The points $P$ and $P'$ are inverse points with respect to circle $C$. And if $P'$ is the inverse of $P$, then $P$ is also the inverse of $P'$. The inversion has interchanged the inside and the outside of circle $C$, leaving only the points on the circle fixed. Thus, if a moving point $P$ approaches the center $O$, its image $P'$ will recede into the distance, resulting in point $O$ being the point at infinity under the inversion. Also, since each point within the plane corresponds to one and only one point in its image, a singular *inverse function* is possible.

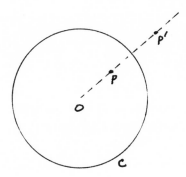

An *inverse function* in mathematics is the function that replaces another function when the dependent and independent variables of the first function are interchanged for an appropriate set of values of the dependent variable. Take the expression $x^2 + 3x - 6$. This has no definite value until the value of $x$ is assigned. Thus, the value of this expression is a *function* of the value of $x$: $f(x) = x^2 + 3x - 6$. If $x$ is assigned the value of 2, then $f(2) = 2^2 + (3 \times 2) - 6 = 4$.

If a function $f(x)$ is known, it is possible to attempt to solve the equation $u = f(x)$ for $x$, so that $x$ will appear as a function of $u$: $x = g(u)$. The function $g(u)$, then, is called an *inverse function* of $f(x)$. This will lead to a unique result only if $u = f(x)$ defines a unique mapping of $x$ into $u$ (as in the circular reflections). For example, the function $m = 2n$, where $n$ is limited to the set $S$ of integers and $m$ to the set $T$ of even integers, does lead to a one-to-one correspondence, and the inverse function $n = m/2$ is uniquely defined.

"I can see all of it when I get upon a chair—all but the bit just behind the fireplace. Oh! I do so wish I could see *that* bit! I want so much to know whether they've a fire in the winter: you never *can* tell, you know, unless our fire smokes, and then smoke comes up in that room too—but that may be only pretence, just to make it look as if they had a fire. Well then, the books are something like our books, only the words go the wrong way: I know *that,* because I've held up one of our books to the glass, and then they hold up one in the other room." [Carroll, *Through the Looking-Glass,* p. 135–136]

Mirrors not only invert images, they reflect images. The mathematical formula for a reflected image is $\dfrac{1}{p} + \dfrac{1}{q} = \dfrac{2}{R}$, where $p$ equals the distance the object is from the center of the mirror the reflected image crosses the line from the center of the mirror to the object, and $R$ equals the radius of the circle.

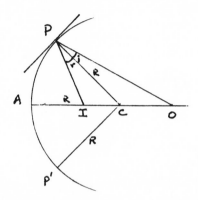

In the above diagram $PAP'$ is a cross section of a spherical mirror. $O$ is an object from which light rays are emitted. $OP$ is a light ray that hits the mirror at $P$. The light is reflected so that the angle of incidence $i$ equals the angle of reflection $r$. $C$ is the center of the spherical surface, i.e., the center of $PAP'$. $CP$ is perpendicular to the circle, i.e., perpendicular to the tangent at $P$. $F$ is the focus of the mirror; $p$ equals the distance $OA$; $q$ equals the distance $IA$ ($I$ is where the reflected ray crosses $AO$). $R$ equals $CA$, the radius of the circle.

The above diagram is of a concave mirror. Concave mirrors produce *real* images. If a concave mirror has a parabolic surface, it will produce a sharp image. Such mirrors are used for telescopes, searchlights, and lighthouses. The large astronomical mirrors used in the great reflecting telescopes at Mount Palomar and Mount Wilson have been made not to deviate from a mathematically perfect paraboloid of revolution by more than one millionth of an inch, and the final polishing process took over three years.

Flat and convex mirrors do not produce a real focused image. If the mirror is flat, the image will appear as far into the mirror as the object is in front of the mirror: i.e., the radius $R$, becomes infinite, thus, $\dfrac{1}{p} + \dfrac{1}{q} = 0$, or $p = -q$. A convex mirror will produce a virtual image behind itself at a distance ever closer to the mirror as the curvature increases.

Prior to the thirteenth century mirrors were made of polished gold, steel, or bronze. Glass mirrors with reflecting coatings appeared in the thirteenth century, and for the next two centuries Venice had a monopoly on them. After 1840 silvered mirrors were manufactured, and evaporated aluminum came into use in 1932. Most commercial mirrors are now made by a chemical process in which the silver from silver nitrate is deposited on glass and then coated by shellac or paint to keep the air from tarnishing it. The front of the glass surface produces a faint second image, since glass reflects at about 5 percent. The main reflection of a new silvered mirror is about 95 percent.

Binoculars are sometimes constructed from four plane mirrors. However, the most popular toy made of multiple mirrors is the kaleidoscope. An ordinary kaleidoscope is constructed from two plane mirrors inclined at $\pi/3$ or $\pi/4$ with an object (or objects) placed in the angle between them so that it is reflected by both. This causes the object to be reproduced six or eight times (depending on the angle), creating a symmetrical arrangement.

If two mirrors are connected at an edge so that they can be swung together or apart as a door might be opened or closed, then the angle between them, $\pi/2$, produces, in addition to the object itself, $2n$ images. If the mirrors are parallel, a theoretically infinite number of images is produced.

If the two mirrors are vertical, and a third vertical mirror is introduced, each pair of the three mirrors can be placed to form an angle of $\pi/n$. Thus, any horizontal section will be a triangle of angles $\pi/l$, $\pi/m$, and $\pi/n$, where $l$, $m$, and $n$ are integers, producing the following equation: $\frac{1}{l} + \frac{1}{m} + \frac{1}{n} = 1$, the solutions of which are 3, 3, 3; 2, 3, 6; and 2, 4, 4. In every case, the number of images is infinite. Various isogonal tessellations (equal-angular mosaic patterns) are obtained by placing a point-object in various positions in the triangle (a lighted candle placed between three mirrors produces an amazing effect). If the point-object is placed at a vertex of the triangle, where an angle-bisector meets the opposite side, or at the in-center (where all three angle-bisectors meet), the tessellations are regular polygons.

If the third mirror is placed horizontally instead of vertically (i.e., if the other two mirrors "stand" on it), the number of images is $4n$, where $\pi/n$ equals the angle between the two vertical mirrors (thus, the number of images is no longer infinite).

Four mirrors produce solid tessellations. If the mirrors are placed on planes inclined at angles that are submultiples of $\pi$, tetrahedra of three different shapes can be formed.

Five mirrors arranged in the form of certain triangular prisms produce solid tessellations of prisms.

Six mirrors arranged rectangularly in three pairs of parallels (as in an ordinary room) produce a solid tessellation of rectangular blocks. A kaleidoscope cannot have more than six mirrors.

There was a book lying near Alice on the table, and while she sat watching the White King (for she was still a little anxious about him, and had the ink all ready to throw over him, in case he fainted again), she turned over the leaves, to find some part that she could read, "—for it's all in some language I don't know," she said to herself.

It was like this.

JABBERWOCKY                              ЈABBEᴙᴡОСКУ

'Twas brillig, and the slithy toves          'Twas brillig, and the slithy toves
  Did gyre and gimble in the wabe:          Did gyre and gimble in the wabe:
All mimsy were the borogoves,                All mimsy were the borogoves,
  And the mome raths outgrabe.            And the mome raths outgrabe.

[Carroll, *Through the Looking-Glass,* p. 153]

Mirror writing (as in the above example) is simply the extension of the properties of the mirror into writing. A related activity is the palindrome (a word, line, verse, etc., that reads the same forward and backward, i.e., Madam, I'm Adam). This requires a symmetry not necessary in mirror writing. For instance, Lewis Carroll's interesting discovery that evil spells live in mirror writing does not make it a palindrome.

Mirror writing is also a form of writing called "inversions" by Scott Kim (the same activity is referred to as "designatures" by Scott Morris). This form of writing explores the various symmetrical possibilities of what might be called the geometrical properties of letters, words, and phrases, that is, how they appear on the page (e.g., upside down writing). It is related to the rebus (the representation of a word or phrase with pictures or symbols) and emblematic poetry (poetry that looks like the subject of the poem, e.g., a poem about Jesus written in the shape of a cross).

As Martin Gardner points out, Lewis Carroll's use of inversion extends into the humor of logical contradiction. Gardner's list of examples includes: The Red Queen, who knows of a hill so large that, compared to it, this hill is a valley; the eating of dry biscuits to quench thirst; the messenger who whispers by shouting; and Alice running as fast as she can to stay in the same place. Such logical contradictions are the essence of mathematical paradoxes. (Refer to Paradoxes of the Infinite and Zeno's Paradoxes for discussions of some of the logical contradictions that lie at the center of mathematics.)

And now, one final quotation from *Through the Looking-Glass* (p. 153):

"Oh, Kitty, how nice it would be if we could only get through into Looking-glass House! I'm sure it's got, oh! such beautiful things in it! Let's pretend there's a way of getting through into it, somehow, Kitty. Let's pretend the glass has got all soft like gauze, so that we can get through. Why, it's turning into a sort of mist now, I declare! It'll be easy enough to get through—"

BIBLIOGRAPHY

Averbach, Bonnie, and Orin Chein. *Mathematics: Problem Solving Through Recreational Mathematics.* San Francisco: W. H. Freeman and Co., 1980.

Bakst, Aaron. *Mathematics: Its Magic and Mastery.* New York: D. Van Nostrand Co., 1941.

Ball, W. W. Rouse. *Mathematical Recreations and Essays.* 1892. Rev. by H. S. M. Coxeter. London: Macmillan and Co., 1939.

Bergerson, Howard W. *Palindromes and Anagrams.* New York: Dover, 1973.

Cadwell, J. H. *Topics in Recreational Mathematics.* London: Cambridge University Press, 1970.

Carroll, Lewis. *Through the Looking-Glass: And What Alice Found There.* 1896. Reprinted in *The Complete Works of Lewis Carroll.* New York: Random House, n.d.

Courant, Richard, and Herbert Robbins. *What Is Mathematics?: An Elementary Approach to Ideas and Methods.* New York: Oxford University Press, 1941.

Gardner, Martin, ed. "Notes." In *The Annotated Alice.* New York: Bramhall House, 1960.

Greenstein, Jesse L. "Mirror." In *The American People's Encyclopedia.* Ed. by Walter Dill Scott, Franklin J. Meine, and W. Stewart Wallace. Chicago: Spencer Press, 1954.

Kim, Scott. *Inversions: A Catalog of Calligraphic Cartwheels.* Peterborough, N.H.: BYTE Books, McGraw Hill, 1981.

Kline, Morris. *Mathematical Thought from Ancient to Modern Times.* New York: Oxford University Press, 1972.

――――. *Mathematics and the Physical World.* New York: Thomas Y. Crowell Co., 1959.

**MIRROR WRITING.** See MIRRORS.

**MÖBIUS CHESS.** See FAIRY CHESS.

**THE MÖBIUS STRIP** reigns as the most popular of the one-sided surfaces common in topology (the contemporary branch of mathematics concerned with properties that remain invariant when a structure is given continuous deformation).

Imagine a magician taking a strip of paper, folding it once (see the illustration below), connecting the ends, and cutting it lengthwise.

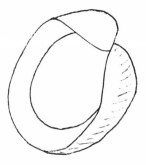

Will the result be two separate rings or two intertwined rings? The answer is neither, and magicians have amazed their audiences with this for centuries. What, then, is the answer? The result will be one ring with double the diameter of the original ring.

Mathematicians usually begin a discussion of the Möbius strip by asking if it is possible for a sheet of paper to have a single edge and only one side, so that an ant could crawl from any point on the paper to any other point without crossing an edge. After allowing the unsuspecting initiate suitable time to struggle with the problem, the mathematician will then form a Möbius strip and have the initiate start a line at any point and continue it until it returns to itself. This produces a line on both sides of the strip (really only on one side, as there really is only one side).

Once the basic idea is understood, it seems so simple. Why, then, was it not until the nineteenth century that mathematicians first discovered the one-sided surface? Perhaps that question can be answered in the same way similar questions are answered. Mathematicians were not given the freedom to consider such surfaces until an absolute was broken, in this case, the absolute rightness of Euclidean geometry.

Though other mathematicians had struggled with certain postulates of Euclidean geometry (more often to find proof for their seeming rightness than to disprove them) for some 2,000 years, it was not until 1829 that Nikolas Ivanovich Lobatchewsky published Plangeometry, a geometry which, containing no contradictions, is as logical as Euclid's (*see* Euclid's Theory of Parallels). This opened the door for non-Euclidean geometry and what was to become topology.

August Ferdinand Möbius (1790–1868), a student of Karl Friedrich Gauss (who is usually given credit for first conceiving a geometry independent of Euclid's), and J. B. Listing both came up with the idea of what has since been called the Möbius strip in 1858.

This strip holds numerous surprises. Martin Gardner indicates a number of them in his article, "Mathematical Games: Curious Figures Descended from the Moebius Band, Which Has Only One Side and One Edge." Consider, for instance, the following: Two strips of paper are placed one on top of the other, given a single half twist, and joined at the ends, all as if they were one strip, as in the following:

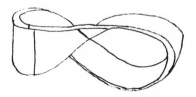

The strips are obviously separate. An ant could walk between the strips forever, its feet on one, its back rubbing the other. However, if it made a mark with its foot on the floor, and circled around the strips until it returned to that mark, the mark would appear on the ceiling. If it circled around again, the mark would appear back on the floor. What appears to be two bands is actually one large band!

Richard Courant and Herbert Robbins, *What Is Mathematics?*, described a "cross-cap," formed through the deformation of a strip which is allowed to intersect itself, as in the following illustration:

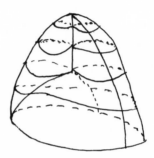

Dr. Bryant Tuckermann stretched the edge of a Möbius strip until it became a triangle, while the strip remained free from intersections, as in the following illustration:

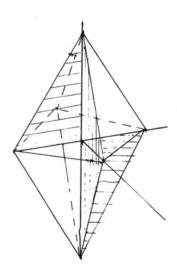

A related one-sided surface is the Klein bottle, which has a closed surface, but no inside or outside. It was invented by Felix Klein in 1882.

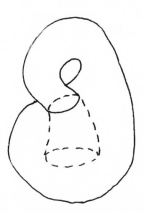

The Klein bottle, or punctured Klein bottle, has only one edge, the edge formed by the hole cut in its side. It differs from the Möbius strip, however, in that it has a loop number (the number of loops—closed curves cut on a surface that do not cross over the surface—that can be cut on a surface without dividing it into two or more pieces) of two; the Möbius strip has a loop number of one.

Luc Etienne, a member of the Oulipo (Ouvroir de Littérature Potentielle— Workshop of Potential Literature) developed a form of poetry based on the Möbius strip. Lines of a poem are interwoven by writing half of the poem on one side of the strip and the other half on the other side of it (upside down). The strip is then twisted and joined at the ends.

For similar play refer to Geometric Dissections; Geometric Problems and Puzzles; Mazes and Labyrinths; Tangrams.

BIBLIOGRAPHY

Armstrong, James W. *Elements of Mathematics*. London: Macmillan and Co., 1970.
Ball, W. W. Rouse. *Mathematical Recreations and Essays*. 1892. Rev. by H.S.M. Coxeter. London: Macmillan and Co., 1939.
———. *A Short Account of the History of Mathematics*. 1888. Reprint. London: Macmillan and Co., 1935.
Behnke, H., F. Bachmann, K. Fladt, and H. Kunle. *Fundamentals of Mathematics. Vol. II: Geometry*. Trans. by S. H. Gould. Cambridge, Mass.: MIT Press, 1974.
Courant, Richard, and Herbert Robbins. *What Is Mathematics?* New York: Oxford University Press, 1941.
Coxeter, H.S.M. *Non-Euclidean Geometry*. Toronto: University of Toronto Press, 1942.
Gardner, Martin. "Mathematical Games: Curious Figures Descended from the Moebius

Band, Which Has Only One Side and One Edge.'' *Scientific American* 196, no. 6 (June 1957): 166–172.

————. ''Mathematical Games: The Flip-Strip Sonnet, the Lipogram and Other Mad Modes of Wordplay.'' *Scientific American* 236, no. 2 (Feb. 1977): 121–126.

————. ''Mathematical Games: The World of the Möbius Strip: Endless, Edgeless and One-sided.'' *Scientific American* 219, no. 6 (Dec. 1968): 112–115.

Hering, Carl. ''A Flat Band with Only One Surface and One Edge.'' *Scientific American* 110, no. 8 (Feb. 21, 1914): 156.

Kline, Morris. *Mathematical Thought from Ancient to Modern Times.* New York: Oxford University Press, 1972.

Meserve, Bruce E. *Fundamental Concepts of Geometry.* London: Addison-Wesley Publishing Co., 1955.

Newman, James R. *Volume One of The World of Mathematics: A Small Library of the Literature of Mathematics from A'h-mosé the Scribe to Albert Einstein, Presented with Commentaries and Notes by James R. Newman.* New York: Simon & Schuster, 1956.

Northrop, Eugene P. *Riddles in Mathematics: A Book of Paradoxes.* New York: D. Van Nostrand Co., 1944.

Steinhaus, H. *Mathematical Snapshots.* New York: Oxford University Press, 1969.

Tietze, Heinrich. *Famous Problems of Mathematics: Solved and Unsolved Mathematical Problems from Antiquity to Modern Times.* Trans. by Beatrice Devitt Hofstadter and Horace Komm. New York: Graylock Press, 1965.

Wolfe, Harold E. *Introduction to Non-Euclidean Geometry.* New York: Holt, Rinehart & Winston, 1945.

**MONEY PROBLEMS.** See ARITHMETIC AND ALGEBRAIC PROBLEMS AND PUZZLES.

**MOVING ROWS.** See ZENO'S PARADOXES.

**MUGGINS.** See DOMINOES.

**MULTIGRADES.** See NUMBER PATTERNS, TRICKS, AND CURIOSITIES.

**NARCISSISTIC NUMBERS.** See NUMBER PATTERNS, TRICKS, AND CURIOSITIES.

**NASIK SQUARES.** See MAGIC SQUARES.

**NECKTIE PARADOX.** See FALLACIES.

**NIM** is a two-person number game that includes such variations as Pebbles, Odds, Tac Tix (Bulo), Wythoff's Game, Hackenbush, Mancala, and Nymphabet.

The general rules of Nim are simple. A number of objects are divided arbitrarily into several piles. Two players alternate, selecting any number of objects from any one of the piles. At least one object must be selected in each turn. Whoever picks up the final object wins (or loses).

The game's origin is unknown, though Martin Gardner (*The Scientific American Book of Mathematical Puzzles and Diversions,* p. 151), suggests that its origin might be Chinese, possibly because of its similarity to the Chinese game of fan-tan. R. Archibald, "The Binary Scale of Notation, a Russian Peasant Method of Multiplication, the Game of Nim and Cardan's Rings," points out that fan-tan had, in fact, wrongly been used to designate the game that Charles L. Bouton named Nim in 1901.

Martin Gardner, *Wheels, Life and Other Mathematical Amusements,* suggests a number of possible sources for the name. Perhaps Bouton had the German *nimm* (the imperative of *nehmen,* "to take") in mind, or perhaps he got it from the archaic English *nim* ("take"), which became a slang word for "steal." Other possible sources are John Gay's *Beggar's Opera* (1727), where a snuffbox is "nimm'd by Filch," or William Shakespeare's Corporal Nym, a thieving attendant of Falstaff in *The Merry Wives of Windsor*. A final possibility is that Nim may simply be the word "win" inverted.

Whatever its origin, the game seems to have a universal appeal. A form of it known as Pebbles or Odds has been played in Africa and Asia for centuries. In this version, an odd number of pebbles, seeds, or whatever is placed in a pile, and players take turns selecting one, two, or three until all have been drawn. The player with an odd number in possession wins.

Gyles Brandreth, *Indoor Games* and *The World's Best Indoor Games,* describes a number of matchstick games based on Nim. In the basic version, fifteen matchsticks are laid out in rows of seven, five, and three.

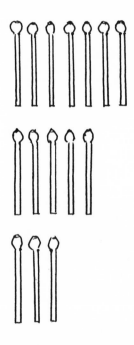

Each player, in turn, may pick up as many matches from many single row as he wishes. Depending on the rules chosen by the players, the object is to pick up the final match or not to pick up the final match (usually the latter). Brandreth offers two variations which he calls One Line Nim and Nim Twenty One. In the former, matchsticks are laid out in a straight line, and players take turns picking up one, two, or three matches; in the latter, twenty-one matches are laid out in a single row and the same process is followed.

Commonly, Nim is played with twelve pennies placed in rows of three, four, and five.

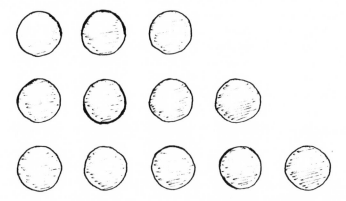

A good player soon realizes that he can always win if he can leave two rows with more than one penny in each and the same number in each, or if the move leaves one penny in one row, two pennies in the second, and three pennies in the third.

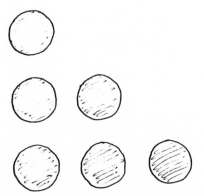

This is not such an amazing discovery, but at the turn of the century it was found that this winning strategy could be generalized and used for any number of rows with any number of objects in each. The solution is based on simple binary theory. Charles L. Bouton ("Nim, a Game with a Complete Mathematical Theory"), in addition to giving the game its most commonly used name, published the first full proof of this.

For anyone familiar with binary mathematics, the solution is simple. In Bouton's terminology, every combination of objects is either "safe" or "unsafe." The combination is "safe," i.e., guaranteeing a win, if the binary notation for each column adds up to zero or an even number. In the version of Nim involving three, four, and five penny rows, the binary notation is as follows:

Binary notation (i.e., numbers written as digits "0" or "1"): 3 pennies = 11; 4 pennies = 100; 5 pennies = 101. Therefore, the following table can be set up:

| Binary Notation: | 4 2 1 |
|---|---|
| Stack of 3 pennies | 1 1 |
| Stack of 4 pennies | 1 0 0 |
| Stack of 5 pennies | 1 0 1 |
| Totals | 2 1 2 |

The middle column adds up to 1, an "unsafe" position.

The only move that will change this to a "safe" position is to remove two pennies from the top pile (the stack of three pennies), thus changing it to a 1, and the total to 202 (all even numbers):

| Binary Notation: | 4 2 1 |
|---|---|
| Stack of 3 pennies (now 1) | 1 |
| Stack of 4 pennies | 1 0 0 |
| Stack of 5 pennies | 1 0 1 |
| Totals | 2 0 2 |

In the reverse game (where a player wins by forcing his opponent to take the final marker), the player wins by leaving an odd number of one-unit stacks. For example, one stack with one object and two stacks with no objects places him in a winning position, i.e., the other player is forced to choose the final object.

Obviously, games of Nim involving larger numbers of objects demand considerable ability to perform calculations in a binary system, and if pencil and paper are not allowed, the player first able to perform the necessary calculations mentally will be the winner.

Since binary mathematics lies at the foundation of the modern explosion in computers, it goes without question that Nim-playing machines are possible. Edward U. Condon and an associate invented the first such machine in 1940. It was patented as the Nimatron, built by the Westinghouse Electric Corporation, and exhibited at the New York World's Fair, playing 100,000 games and winning 90,000. In 1941, Raymond M. Redheffer designed a more efficient model. In 1951, Nimrod, a Nim-playing robot, was exhibited at the Festival of Berlin and later at the Berlin Trade Fair, where it defeated the economics minister, Dr. Erhard, in three straight games.

In the mid-1940s Piet Hein of Copenhagen developed a version of Nim called Tac Tix (or Bulo) in which the objects are arranged in a square similar to Samuel Loyd's famous Fifteen Puzzle.

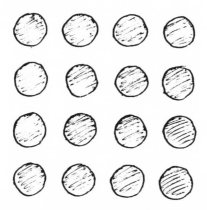

Once again, the players alternately remove objects from one row at a time. However, in this version the row may be *either* a vertical or a horizontal row, and objects removed must be adjoining. Tac Tix must be played in reverse, what Martin Gardner calls the misère form (the player taking the final counter losing), because of an easy strategy for the straight form. In the straight form, if the square contains an odd number of objects on each side, the first player can win by taking the center object and playing symmetrically against his opponent. On the other hand, in the straight game, if the square contains an even number of objects on each side, the second player need only play symmetrically to win. The reverse or misère form cannot be played this way.

Hein has applied the principle of intersecting sets of counters to other two- and three-dimensional patterns, e.g., the vertices and intersections of pentagrams and hexagrams.

Eliakim H. Moore, "A Generalization of the Game Called Nim," analyzed a form of Nim where the players are allowed to take from any number of rows any number of objects not to exceed a designated number "x." In this version the same binary rules apply, a "safe" position being defined as one in which each column of binary numbers totals a number evenly divisible by $x + 1$.

| Designated Number = 1 | |
|---|---|
| Binary Notation: | 4 2 1 |
| Stacks of 3 pennies | 1 1 |
| Stacks of 4 pennies | 1 0 0 |
| Stacks of 5 pennies | 1 0 1 |
| Totals | 2 1 2 |

Here $x + 1 = 2$. Therefore, the center cannot be evenly divided by $x + 1$, and the position is "unsafe."

Gardner, *Wheels, Life and Other Mathematical Amusements*, discusses a form of Nim played on a checkerboard. The counters (an equal set for each player, no fewer than two for each) are set up across from each other randomly:

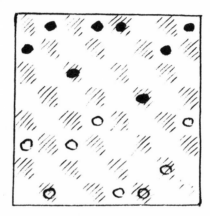

Players, in turn, move one of their markers forward in its column. The marker may not move onto or past the opponent's marker in the same column. When the two markers in any column meet, neither may be moved again. The last player to move wins. The game may be complicated by allowing the players to move backward as well as forward. A backward move would be equivalent to adding counters to a Nim row.

Gardner, *Wheels, Life and Other Mathematical Amusements*, also discusses Hackenbush (also called Graph and Chopper, and Lizzie Borden's Nim), a game based on Nim developed by John Horton Conway. In Hackenbush, a pattern of disconnected graphs is drawn on a sheet of paper, as in the figure below:

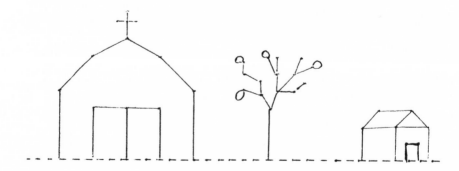

The following terms are used: An "edge" is any line joining two "nodes" (points) or one node to itself, in which case the edge is a "loop." There may be more than one edge between two nodes. Each graph rests upon a base line, which is not a part of the graph. Nodes on the base line are called base nodes.

Two players in turn remove an edge at a time. If the edge removed also separates the rest of the graph or any portion of the graph from a continuous, unbroken path of edges to the base line, that graph or portion of the graph is also eliminated. The player taking the final edge wins.

Each picture is either safe or unsafe and, as in Nim, the winning strategy is to change every unsafe pattern to a safe one. The way to determine this is to measure the "weight" of each graph in the picture. To do this, first collapse all of the "cycles" (closed circuits of two or more edges) to loops, as in the figure below:

Next turn each loop into a single edge:

Now calculate the graph's weight. Label all the edges with a terminal node (a node unconnected to another edge) as a 1. Label all edges that support only one additional edge as a 2. Label the remaining edges with one more than the Nim sum of all the edges it immediately supports, as in the figure below:

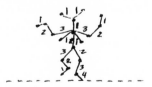

The Nim sum (often called the "Grundy" or "Sprague-Grundy" number, after Roland Sprague and P. M. Grundy, who separately came up with the general theory of take-away games by assigning single numbers to each state of the game) is determined by expressing each graph or the total picture as the sum of distinct powers of 2, eliminating all pairs of like powers, and adding the powers that remain. For example, in the game of Nim played with stacks of three, four, and five pennies, the following Nim sum of 2 results:

$3 = 2 + 1$
$4 = 4$
$5 = 4 + 1$

The 4's and the 1's cancel each other out, leaving a Nim sum of 2. As shown earlier, where the columns in binary notation added up to 212, this is an unsafe position. It can be made safe by removing two objects from the stack of three, thus changing the Nim sum to zero (a safe position) and the binary column totals to 202 (all even numbers—a safe position).

If Hackenbush is played with only one graph with only one base node, the first player simply takes the base node, collapsing the graph and winning. If the game is played with one graph with two base nodes, as in the door in the house above, the object is to reduce the Nim sum to zero (a safe position), which in the case of the door is done by removing the door's top edge.

If the picture consists of more than one graph, the graphs are treated as if they were separate stacks in Nim. If the Nim sum of all the weights is zero, the picture is safe and the second player is assured of winning.

In 1907, W. A. Wythoff presented a variation on the standard game of Nim, and in 1953, H.S.M. Coxeter analyzed and expressed this variation in a closed form. According to John C. Holliday, "Some Generalizations of Wythoff's Game and Other Related Games," it has "probably been a major stimulus for the study of complementary sequences." According to Coxeter, "The Golden Section, Phyllotaxis, and Wythoff's Game," Wythoff's Game is simply another application of the golden section (the division of a line segment into two unequal parts such that the ratio of the whole to the large part is equal to the ratio of the larger to the smaller).

In Wythoff's Game, two stacks of counters are placed on a table. A player either removes an arbitrary number of counters from one stack or an equal number from both stacks. The player leaving no counters after his move wins the game. Once again, a player can determine the outcome by continually leaving "safe" combinations, which can be done by always making sure that the stack with the smallest number has the smallest natural number not already used, and that the stack with the larger number is chosen so that the difference of the two numbers in the $n$th pair is $n$, i.e., (1,2), (3,5), (4,7), (6,10), (8,13), (9,15), (11,18), (10,20). . . .

David Parlett, *Botticelli and Beyond: Over 100 of the World's Best Word Games,* created Nymphabet, an alphabetical version on Nim. In Nymphabet, the twenty-six letters of the alphabet serve as the original group to be eliminated.

Each player in turn writes a word. The first word must begin with A, which is then crossed off the alphabet. If the word contains a B after the A, the B is also crossed off the alphabet. The same is done for each letter of the alphabet in succession:

A̲bout
Both A and B are crossed off the alphabet.

The next player must write a word that begins with the letter of the alphabet next in line that has not been crossed off:

C̲ard̲
The C and the D are crossed off the alphabet.

This continues until a player is forced to use the Z. Parlett suggests that the game is most interesting if the writer of the final word loses.

The game can be varied by treating the alphabet as a circle, A following Z, and beginning at any point in the alphabet. It can also be varied by giving points for each new letter used, and either adding or multiplying the points to determine the final score.

For similar mathematical recreations refer to Checkers; Chess; Dominoes; Tictactoe.


BIBLIOGRAPHY

Archibald, R. C. "The Binary Scale of Notation, a Russian Peasant Method of Multiplication, the Game of Nim, and Cardan's Rings." *American Mathematical Monthly* 25 (March 1918): 139–142.

Bouton, Charles L. "Nim, a Game with a Complete Mathematical Theory." *Annals of Mathematics,* Ser. 2, Vol. 3 (1901–1902): 35–39.

Brandreth, Gyles. *Indoor Games.* London: Hodder & Stoughton, 1977.

———. *The World's Best Indoor Games.* New York: Pantheon Books, 1981.

Brooke, Maxey. *Fun for the Money: Puzzles and Games with Coins.* New York: Charles Scribner's Sons, 1963.

Coxeter, H.S.M. "The Golden Section, Phyllotaxis, and Wythoff's Game." *Scripta Mathematica* 19 (1953): 135–43.

Gardner, Martin. "Mathematical Games: Concerning the Game of Nim and Its Mathematical Analysis." *Scientific American* 198 (Feb. 1958): 104–111.

———. *The "Scientific American" Book of Mathematical Puzzles and Diversions.* New York: Simon & Schuster, 1959.

———. *Wheels, Life and Other Mathematical Amusements.* New York: W. H. Freeman and Co., 1983.

Hemmings, Ray. "The Amazing Dr. Nim." *Mathematics Teaching,* No. 40 (Autumn 1967): 44.

Holladay, John C. "Some Generalizations of Wythoff's Game and Other Related Games." *Mathematics Magazine* 41 (Jan. 1968): 7–13.

Kraitchik, Maurice. *Mathematical Recreations*. 2d rev. ed. New York: Dover, 1953.

Moore, Eliakim H. "A Generalization of the Game Called Nim." *Annals of Mathematics,* Ser. 2, Vol. 11 (1910): 93–94.

Parlett, David. *Botticelli and Beyond: Over 100 of the World's Best Word Games*. New York: Pantheon Books, 1981.

Schaaf, W. L. "Number Games and Other Mathematical Recreations." In *Macropaedia: Encyclopaedia Britannica,* 1985.

**NIM TWENTY ONE.** See NIM.

**NON-EUCLIDEAN GEOMETRY.** See EUCLID'S THEORY OF PARALLELS.

**NOUGHTS AND CROSSES.** See TICTACTOE.

**NUMBER GUESSING.** See GUESSING NUMBERS.

**NUMBER MYSTICISM.** See NUMEROLOGY.

**NUMBER PATTERNS, TRICKS, AND CURIOSITIES** are those mathematical oddities and amusing relationships that are found in the arithmetic and algebraic manipulations of numbers.

For example, there are several natural numbers which, when subjected to simple arithmetical operations, obey the *law of series*:

If 37 is multiplied by 3, or any multiple of 3 up to 27, the product is three figures of the same kind in sequence:

$37 \times 3 \ = 111$
$37 \times 6 \ = 222$
$37 \times 9 \ = 333$
$37 \times 12 = 444$
and so on . . .

If 37 is multiplied by multiples of 3 higher than 27, the first and last digits continue the sequential pattern:

$37 \times 30 = 1,110 \ (10)$
$37 \times 33 = 1,121 \ (11)$
$37 \times 36 = 1,332 \ (12)$
and so on . . .

If 37 is reversed to 73 and multiplied by 3, and its progressive multiples, the final digits of each successive product form a descending sequence:

73 × 3  = 219 (9)
73 × 6  = 438 (8)
73 × 9  = 657 (7)
73 × 12 = 876 (6)
and so on . . .

The following patterns, once again, are the result of natural numbers subjected to the basic operations of arithmetic:

| | | |
|---|---|---|
| $(1)^2$ | = | 1 |
| $(11)^2$ | = | 121 |
| $(111)^2$ | = | 12321 |
| $(1111)^2$ | = | 1234321 |

and so on . . .

$$1 \times 8 + 1 = 9$$
$$12 \times 8 + 2 = 98$$
$$123 \times 8 + 3 = 987$$
$$1234 \times 8 + 4 = 9876$$
$$12345 \times 8 + 5 = 98765$$

and so on . . .

123,456,789 × 45 = 5,555,555,505
987,654,321 × 45 = 44,444,444,445
123,456,789 × 54 = 6,666,666,606
987,654,321 × 54 = 53,333,333,334

15,873 × 7 = 111,111
31,746 × 7 = 222,222
47,619 × 7 = 333,333
63,492 × 7 = 444,444
79,365 × 7 = 555,555
95,238 × 7 = 666,666
111,111 × 7 = 777,777
126,948 × 7 = 888,888
142,857 × 7 = 999,999

$142,857 \times 7 = 999,999 / 9 = 111,111$

$285,714 \times 7 = 1,999,998 / 9 = 222,222$

$428,571 \times 7 = 2,999,997 / 9 = 333,333$

$571,428 \times 7 = 3,999,996 / 9 = 444,444$

$714,285 \times 7 = 4,999,995 / 9 = 555,555$

$857,142 \times 7 = 5,999,994 / 9 = 666,666$

In the preceding example, the numbers being multiplied by 7 contain the same sequence of digits, starting at a different point in the sequence each time.

$$0 \times 9 + 1 = 1$$
$$1 \times 9 + 2 = 11$$
$$12 \times 9 + 3 = 111$$
$$123 \times 9 + 4 = 1111$$
$$1234 \times 9 + 5 = 11111$$

and so on . . .

$$9 \times 9 + 7 = 88$$
$$9 \times 98 + 6 = 888$$
$$9 \times 987 + 5 = 8888$$
$$9 \times 9876 + 4 = 88888$$

and so on . . .

$1 \times 1 \times 1 = 1$

$2 \times 2 \times 2 = 3 + 5$

$3 \times 3 \times 3 = 7 + 9 + 11$

$4 \times 4 \times 4 = 13 + 15 + 17 + 19$

$5 \times 5 \times 5 = 21 + 23 + 25 + 27 + 29$

and so on . . .

$2^2 = 1 \ + 2 + 1 \ = 4$

$3^2 = 4 \ + 2 + 3 \ = 9$

$4^2 = 9 \ + 2 + 5 \ = 16$

$5^2 = 16 + 2 + 7 \ = 25$

$6^2 = 25 + 2 + 9 \ = 36$

$7^2 = 36 + 2 + 11 = 49$

$8^2 = 49 + 2 + 13 = 64$

$9^2 = 64 + 2 + 15 = 81$

and so on . . .

In the preceding example, the number for each successive square equals the number of the previous square plus 2 plus a successive odd number.

When numbers are linked to each other in some form of logical sequence, they form a series or chain, and the *law of series* is the rule or set of rules that explains the connection(s). The following series or chain allows the summing of a sequence of numbers that begins with 1:

1 (term or number in the series) summed     $= 1$ or $\dfrac{1 \times 2}{2}$

2 (terms or numbers in the series) summed $= 3$ or $\dfrac{2 \times 3}{2}$

3 (terms or numbers in the series) summed $= 6$ or $\dfrac{3 \times 4}{2}$

4 (terms or numbers in the series) summed $= 10$ or $\dfrac{4 \times 5}{2}$

and so on . . .

The law of series or rule for this series is: $\dfrac{n(n + 1)}{2}$. In other words, any series that begins with 1 and works its way sequentially up to $n$, if summed, will total in accordance with the law of series. Thus, if the sum of all of the numbers up to 100 is desired, rather than adding the first one hundred numbers together one at a time, it is simply necessary to plug 100 into the $n$ slot in the notation:

$$\frac{100 \ (100 + 1)}{2} = 5{,}050$$

If the sequence of numbers does not begin at 1, a more general rule is needed. Notice the following:
A one-digit sequence:

$1 = 1 \times 1 + 0 = 1$
$2 = 1 \times 2 + 0 = 2$
$3 = 1 \times 3 + 0 = 3$
$4 = 1 \times 4 + 0 = 4$
and so on . . .

A two-digit sequence:

$1 + 2 = 2 \times 1 + 1 = 3$
$2 + 3 = 2 \times 2 + 1 = 5$
$3 + 4 = 2 \times 3 + 1 = 7$

$4 + 5 = 2 \times 4 + 1 = 9$

and so on . . .

A three-digit sequence:

$1 + 2 + 3 = 3 \times 1 + 3 = 6$
$2 + 3 + 4 = 3 \times 2 + 3 = 9$
$3 + 4 + 5 = 3 \times 3 + 3 = 12$
$4 + 5 + 6 = 3 \times 4 + 3 = 15$

and so on . . .

A four-digit sequence:

$1 + 2 + 3 + 4 = 4 \times 1 + 6 = 10$
$2 + 3 + 4 + 5 = 4 \times 2 + 6 = 14$
$3 + 4 + 5 + 6 = 4 \times 3 + 6 = 18$
$4 + 5 + 6 + 7 = 4 \times 4 + 6 = 22$

and so on . . .

A five-digit sequence:

$1 + 2 + 3 + 4 + 5 = 5 \times 1 + 10 = 15$
$2 + 3 + 4 + 5 + 6 = 5 \times 2 + 10 = 20$
$3 + 4 + 5 + 6 + 7 = 5 \times 3 + 10 = 25$
$4 + 5 + 6 + 7 + 8 = 5 \times 4 + 10 = 30$

and so on . . .

A close study of the sequences reveals that the number of digits in the sequence times the first number of the sequence, plus the number of digits in the previous sequence added to the number that was added to the previous sequence, produces a general formula for adding any number of natural numbers in sequence:

$n \times a + n - 1 + b$

$n$ = the number of numbers in the sequence
$a$ = the first number of the sequence
$b$ = the number added to the previous sequence

A study of the resulting sequence of $b$ reveals a general pattern to the differences between each successive number (i.e., each successive $b$ is one natural number higher than the previous $b$; $1 - 0 = \underline{1}$; $3 - 1 = \underline{2}$; $6 - 3 = \underline{3}$; $10 - 6 = \underline{4}$; and so on:

| Number of digits in sequence | × | first digit in sequence | + | $b$ | + | $n - 1$ |
|---|---|---|---|---|---|---|
| 1 | × | $a$ | + | 0 | + | 0 |
| 2 | × | $a$ | + | 0 | + | 1 |
| 3 | × | $a$ | + | 1 | + | 2 |
| 4 | × | $a$ | + | 3 | + | 3 |
| 5 | × | $a$ | + | 6 | + | 4 |

The sum of the first $n$ terms of *even* numbers is twice that of all numbers, or: $n(n + 1)$. The formula for starting at *any* even number is: $n \times a + 2(n - 1 + b)$. Thus, the sequence $2 + 4 + 6$ equals: $3(3 + 1) = 3 \times 2 + 2(3 - 1 + 1) = 12$. If the sequence begins with a number higher than 2, the first formula cannot be used (e.g., $4 + 6 + 8 \neq 3(3 + 1)$). The second formula, however, can be used (e.g., $4 + 6 + 8 = 3 \times 4 + 2(3 - 1 + 1) = 18$).

The sum of the first $n$ terms of *odd* numbers is $n^2$ (e.g., $1 + 3 + 5 + 7 = 4^2 = 16$). The general formula for starting at *any* odd number also works (e.g., $3 + 5 + 7 + 9 = 4 \times 3 + 2(4 - 1 + 3) = 24$).

In fact, the general formula works for any sequence of addition. Here are a few examples:

A four-number sequence starting at 2 with 3 added to each successive number: $4 \times 2 + 3(4 - 1) = 18 = 2 + 5 + 8 + 11$.
The general formula is the same as that for an even number with the doubling changed to match the difference between successive (i.e., $c$ in the following formula equals the number each successive number is higher than the previous number: $n \times a + c(n - 1 + b))$.

A three-number sequence starting at 7 with 7 added to each successive number: $3 \times 7 + 7(2 + 1) = 42 = 7 + 14 + 21$.

A four-number sequence starting at 11 with 6 added to each successive number: $4 \times 11 + 6(3 + 3) = 11 + 17 + 23 + 29 = 80$.

These types of sequences or chains are referred to as *arithmetical progressions*. They are progressions with a constant increase or decrease.

A *geometric progression* is one in which the *ratio* between any two successive terms is the same (e.g., $1, 3, 9, 27, 81$; $144, 12, 1, \frac{1}{12}, \frac{1}{144}$). Geometric progressions are central to a number of paradoxes that mathematicians still have not solved to the satisfaction of all (refer to Logical Paradoxes and to Zeno's Paradoxes). Here is an example:

A librarian classifies one-half of the library's books in one month. He classifies one-half of the remaining books in the next month. He classifies one-half of the remaining books in the next month. Each month thereafter he classifies one-half as many books as he did the previous month. Does he ever get all of the books classified?

*Triangular numbers* are those numbers that result from the addition of natural numbers:

$$1 = 1$$
$$3 = 1 + 2$$
$$6 = 1 + 2 + 3$$
$$10 = 1 + 2 + 3 + 4$$
$$15 = 1 + 2 + 3 + 4 + 5$$

and so on . . .

This is the same sequence as $b$ in the general formula for an arithmetic progression.

Triangular numbers squared equal:

$$1^2 = 1^2 = 1^3$$
$$3^2 = (1 + 2)^2 = 1^3 + 2^3$$
$$6^2 = (1 + 2 + 3)^2 = 1^3 + 2^3 + 3^3$$

and so on . . .

Triangular numbers added sequentially in pairs result in *square numbers* (natural numbers squared):

$$1 = 0 + 1 = 1 + 0$$
$$4 = 1 + 3 = (1 + 2) + 1$$
$$9 = 3 + 6 = (1 + 2 + 3) + (1 + 2)$$
$$16 = 6 + 10 = (1 + 2 + 3 + 4) + (1 + 2 + 3)$$
$$25 = 10 + 15 = (1 + 2 + 3 + 4 + 5) + (1 + 2 + 3 + 4)$$

and so on . . .

If triangular numbers and square numbers are added in the same manner as the triangular numbers were added above, they equal *pentagonal numbers*:

$$1 = 1$$
$$5 = 1 + 4$$

12 = 3 + 9
22 = 6 + 16
35 = 10 + 25
and so on . . .

If triangular numbers and pentagonal numbers are added in the same way, they equal *hexagonal numbers*:

1  = 1
6  = 1 + 5
15 = 3 + 12
28 = 6 + 22
45 = 10 + 35
and so on . . .

Figures may be formed of all of these numbers, and thus, they are called *figurate numbers*. Figurate numbers were very important to the ancient Greek mathematicians. For the school of Pythagoras, everything could be explained by number (refer to Numerology and Zeno's Paradoxes), and numbers were given various characteristics, among them shapes. Triangular numbers were those numbers that would form triangles when they were visualized as points on a piece of paper (probably stones in the sand for Pythagoras). Square numbers formed squares; pentagonal numbers formed pentagons, and so on.

*Oblong numbers,* the sums of successive even numbers (e.g., 2 + 4 = the oblong number 6), or the product of two successive numbers (e.g., 2 × 3 = the oblong number 6), form oblong networks of points:

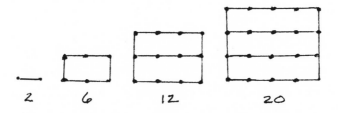

2          6          12          20

A *gnomon* is the part of a parallelogram that is left after a similar parallelogram has been taken away from one of its corners. Gnomons include all of the odd figurate numbers, which in turn can be represented by a right angle:

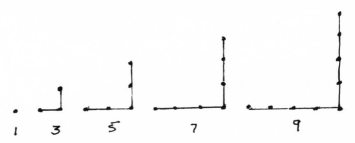

1    3    5    7    9

W. L. Schaaf, "Number Games and Other Mathematical Recreations," suggests that Pythagoras might have come up with his famous theorem ($a^2 + b^2 = c^2$) from contemplating gnomons.

It is possible to add *polygonal number series* (series of numbers which can be represented visually as polygons) to form three-dimensional figurate numbers called *pyramidal numbers*.

The following are polygons formed from triangular, square, pentagonal, and hexagonal numbers:

Triangular numbers:

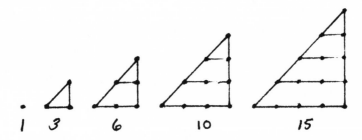

1    3    6    10    15

Square numbers:

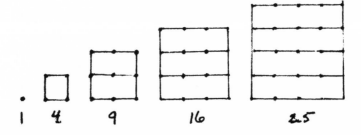

1    4    9    16    25

Pentagonal numbers:

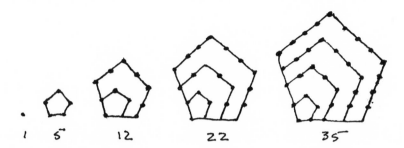

Hexagonal numbers:

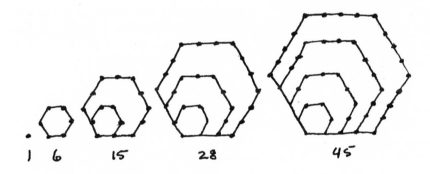

A fascinating sequence of sequences results when each of these sequences is subtracted in sequence from itself:

Triangular:

```
1   3   6   10   15   21   28   36 . . .
  2   3   4    5    6    7    8 . . .
    1   1   1    1    1    1 . . .
```

Square:

```
1   4   9   16   25   36   49   64 . . .
  3   5   7    9   11   13   15 . . .
    2   2   2    2    2    2 . . .
```

Pentagonal;

```
1  5  12  22  35  51  70  91 . . .
  4  7  10  13  16  19  22 . . .
    3  3  3  3  3  3 . . .
```

These larger sequences led to Pascal's triangle, but before detailing that it is worthwhile to consider the *Fibonacci numbers* (as they are also contained within Pascal's triangle).

The Fibonacci numbers are those numbers that make up the sequence formed by adding the two previous numbers to produce the next number in the sequence (i.e., *1, 1, 2, 3, 5, 8, 13, 21 . . .*).

Leonardo of Pisa (c. 1175–c. 1250), better known as Fibonacci, wrote a famous book, *Liber abaci* (1202), which was a defense of the Hindu-Arabic form of notation. Within this book was a trivial problem: Suppose, Fibonacci said, a male-female pair of adult rabbits is placed inside an enclosure to breed. Assume that rabbits start to bear their young in two months after their own birth, producing only a single male-female pair, and that they have one such pair at the end of each subsequent month. If none of the rabbits dies, how many pairs of rabbits will there be inside the enclosure at the end of one year?

This problem is solved through what subsequently became known as the Fibonacci sequence, up to one year's total, i.e., 377.

At the beginning of the nineteenth century, mathematicians began working on sequences of numbers, and E. A. Lucas (1842–1891) was able to generalize the Fibonacci sequence to include *any* two positive integers, each number thereafter being the sum of the two previous numbers. The next simplest sequence has subsequently been named the *Lucas sequence (1, 3, 4, 7, 11, 18 . . .)*.

An amazing property of the Fibonacci series (which also holds for the generalized series) is that the ratio between two consecutive numbers is alternately greater and lesser than the *golden ratio* (thought to be the most pleasing aesthetic proportions). The golden ratio is the ratio between two portions of a line or the dimensions of a plane figure in which the lesser of the two is to the greater as the greater is to the sum of both; A is to B as B is to A + B; approximately 0.618 to 1.000. Furthermore, the ratio between two consecutive integers in the Fibonacci series becomes smaller and smaller as the series continues, the golden ratio serving as the limit to the series.

The golden ratio is $(\phi) = (\sqrt{5} + 1):2 = 1.618 . . .$, which is known as the *golden number*. Its receprical equals 0.618 . . . . Both are related to the roots of $x^2 - x - 1 = 0$, the equation derived from the *divine proportion* of Lucas Pacioli (fifteenth century), which states that $a/b = b/(a + b)$, when $a < b$, by setting $x$

$= b/a$. A *golden rectangle*, then, has a proportion of approximately 8:5 (1.618 . . .) or 3:5 (0.618 . . .):

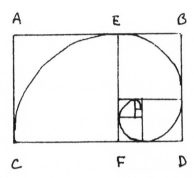

In the above golden rectangle EBDF is also a golden rectangle (created by removing square AEFC). If this process is continued, and circular arcs drawn, the curve formed approximates the logarithmic spiral, a common form found in nature (the logarithmic spiral is a graph of $r = k^\theta$, in polar coordinates, where $k = \phi^{2/\pi}$).

Fibonacci numbers also show up in phyllotaxis, a botanical phenomenon governing the arrangement of petals on many flowers, branches on numerous plants, whorls on pinecones and pineapples, and so on.

*Tribonacci numbers* are those in which the previous three numbers are added to produce the next in the sequence (1, 1, 2, 4, 7, 13, 24, 44, 81 . . .). They were named by Mark Feinberg (*The Fibonacci Quarterly*, Oct. 1963).

*Pascal's triangle* is named after Blaise Pascal (1623–1662), a French mathematician, because he first wrote a treatise on it, *Traité du triangle arithmétique* (Treatise on the Arithmetic Triangle; 1654). The pattern, however, was known long before Pascal's time. Petrus Apianus included it on the title page of an arithmetic book in the early sixteenth century. In 1303 a Chinese mathematician included an illustration of it in a book. Omar Khayyám apparently knew about it as long ago as 1100. Other mathematicians shortly before Pascal's time—Tartaglia (1560), Schenbel (1558), and Bienewitz (1524)—had used the triangle to determine the coefficients in a binomial expansion.

The first twelve rows of the triangle are reproduced below:

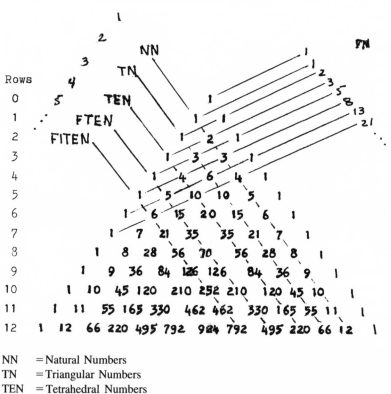

NN    = Natural Numbers
TN    = Triangular Numbers
TEN   = Tetrahedral Numbers
FTEN  = Four Space Tetrahedral Numbers
FITEN = Five Space Tetrahedral Numbers
FN    = Fibonacci Numbers (Sums of the lines)

*Tetrahedral numbers* are cardinal numbers of sets of points that form tetra-hedral arrays (e.g., a triangular pyramid) in three-space.

*Multigrades* are singular identities between sets of numbers and their powers:

($n = 1, 2,$ or $3$)

$$1^n + 5^n + 7^n + 9^n + 12^n + 14^n + 16^n + 20^n = 2^n + 4^n + 6^n + 10^n + 11^n + 15^n + 17^n + 19^n$$

If $n = 1$, then both sides equal 84; if $n = 2$, then both sides equal 1,152; if $n = 3$, then both sides equal 17,766. (This example comes from W. L. Schaaf, "Number Games and Other Mathematical Recreations.")

*Palindromic multigrades* read the same forward and backward. *Prime-number multigrades* consist solely of prime numbers. And so on.

*Pythagorean triples* are any set of positive integers such that $a^2 + b^2 = c^2$. If the three integers are relatively prime (i.e., none of them have a common factor) then the triples form a primitive Pythagorean triple. The only primitive triple that consists of consecutive integers is *3, 4, 5.*

*Amicable or sympathetic numbers* are those pairs of numbers whose divisors equal each other. For example, the divisors of 220 (the number itself is excluded) are *110, 55, 44, 22, 20, 11, 10, 5, 4, 2, and 1.* The divisors for *284* are *142, 71, 4, 2, 1.* If the divisors for *220* are added they equal *284.* If the divisors for *284* are added they equal *220.*

A *perfect number* is a number equal to the sum of all its divisors (not including itself). For example, $6 = 1 + 2 + 3$. Euclid suggested that any number of the form $2^{n-1}(2^n - 1)$ is a perfect number whenever $2^n - 1$ is a prime number. In the eighteenth century Leonhard Euler proved that every even perfect number must be of the form $2^{n-1}(2^n - 1)$, where $2^n - 1$ is a prime number.

Numbers of the form $2^n - 1$ are called *Mersenne numbers,* after the French mathematician Marin Mersenne, who asserted in 1644 that $2^n - 1$ is a prime number for $n \leq 257$ only when $n = 2, 3, 5, 7, 13, 17, 19, 31, 67, 127, and 257.$ The list contains errors (prime numbers for $n = 61$ *or 89 or 107* are omitted; the numbers for $n = 67$ *or 257* are composite). Thus, Mersenne numbers may be *prime numbers* (having no factor other than themselves and 1) or *composite* (composed of two or more prime factors). If $2^n - 1$ is to be a prime number, $n$ must be a prime number. Thus, all even perfect numbers take the form $2^{n-1}(2^2 - 1)$ where both $n$ and $2^{n-1}$ are prime numbers.

Every Mersenne prime corresponds to an even perfect number, and vice versa. All perfect numbers are also triangular. And the sum of the reciprocals of the divisors of a perfect number (including a reciprocal of the number itself) always equal 2 (e.g., $6: \frac{1}{1} + \frac{1}{2} + \frac{1}{3} + \frac{1}{6} = 2$).

Two interesting questions remain about perfect numbers. First, are there any odd perfect numbers? Second, is the number of perfect numbers infinite?

*Fermat numbers,* named after Pierre de Fermat (1601–1665), are numbers of the form $F_n = 2^{2^n} + 1$. Fermat contended that numbers of the form $2^m + 1$ cannot be prime numbers when $m$ contains an odd factor. Furthermore, he stated, if $m$ has no odd factor (i.e., is a power of 2) then $2^m + 1$ is a prime number. Euler proved that Fermat was wrong in claiming that all numbers of this form are prime (i.e., $F_5 = 2^{32} + 1$ is divisible by 641).

*Cyclic numbers* are those numbers that repeat: for example, *142,857* is the generator of the cyclic set *142,857, 428,571, 285,714, 857,142, 571,428, and 714,285* in any scale of notation where the *base* $\geq 9$. If it is in *base 10,* then *142,857* is the cyclic period in the decimal expansion of $\frac{1}{7}$ *(i.e.,* $\frac{1}{7} =$

0.*142857*142857142857 . . .). Successive numbers in the cycle form periods for $\frac{x}{7}$, where $x = 3, 2, 6, 4,$ *and 5* in that order.

*Narcissistic numbers* are those numbers that can be represented by some form of mathematical manipulation of their digits. One form of this is called a *perfect digital invariant*. This is the case when an integer is the sum of $n$th powers of its digits (e.g., $153 = 1^3 + 5^3 + 3^3$). Another form is called a *recurring digital invariant*. This form occurs when the sum of the $n$th powers of the digits of a number equals another number and the sum of the digital powers of that number equals another number, and so on, until the original number is reproduced:

136: $1^3 + 3^3 + 6^3 = 244$

244: $2^3 + 4^3 + 4^3 = 136$

*Automorphic numbers* are numbers whose squares end with the same number. In base 10, only those numbers ending in 5 or 6 apply (e.g., $5^2 = 25$, $25^2 = 625$).

*Strobogrammatic numbers* are numbers that read backward upon a 180° rotation (e.g., 69-96).

*Common or vulgar fractions* are those having integers for both the numerator and the denominator. *Proper fractions* are those where the numerator is less than the denominator. *The Farey series* is the list of all proper common fractions in their lowest terms in order of magnitude up to some arbitrarily assigned limit, say, with a denominator not exceeding 7. In this case the resulting 17 fractions are $\frac{1}{7}, \frac{1}{6}, \frac{1}{5}, \frac{1}{4}, \frac{2}{7}, \frac{1}{3}, \frac{2}{5}, \frac{3}{7}, \frac{1}{2}, \frac{4}{7}, \frac{3}{5}, \frac{2}{3}, \frac{5}{7}, \frac{3}{4}, \frac{4}{5}, \frac{5}{6},$ and $\frac{6}{7}$.

The series was apparently first discovered by C. Haros, but Augustin-Louis Cauchy attributed the series to John Farey, and the name stuck. Farey first observed the series in 1816. It contains some interesting properties. Fractions equidistant from $\frac{1}{2}$ are complementary (their sum equals 1). The number of terms is always odd, and $\frac{1}{2}$ is always the central term. The numerators and denominators for each fraction are obtained by adding the numerators and denominators of the fractions on either side (e.g., $\frac{1}{6} = 1 + \frac{1}{7} + 5$).

A *Diophantine equation* is an algebraic equation with one or more unknowns which have integer coefficients for which integer solutions are sought. The equations may have no solutions, a finite number of solutions, or an infinite number of solutions. A linear Diophantine equation is the simplest: $ax + by = c$.

Martin Gardner, *Aha! Gotcha: Paradoxes to Puzzle and Delight*, offers a paradox involving Diophantine equations called "The Curious Will" (a contemporary version of an age-old problem). Paraphrased, it goes as follows:

A wealthy doctor who owns 11 fancy cars dies. In his final will and testament, he requests that his 11 cars be divided up among his three sons as follows: the oldest is to get half of the cars; the middle son is to get one-fourth of the cars; and the youngest son is to get one-sixth of the cars.

While the three sons attempt unsuccessfully to figure out who gets what, their Aunt Minnie arrives in her own car. She adds her car to the 11, making a total of 12 cars. She then gives 6 cars to the oldest son, 3 cars to the middle son, and 2 cars to the youngest son, satisfying the fractions required. But if the number of cars distributed is added together, a total of 11 is obtained. So Aunt Minnie took back her car.

The paradox works only if the equation $\frac{n}{(n + 1)} = \frac{1}{a} + \frac{1}{b} + \frac{1}{c}$ (n = number of cars) has a solution in positive integers. The reason the paradox occurs is that the original fractions total less than 1 (e.g., there would be some fraction, in this case $\frac{1}{12}$, left over).

The *Pellian equation, $x^2 - Dy^2 = 1$,* was thought up by Pierre de Fermat, who challenged J. Wallis to calculate values of it beyond the first 150 values of $D$ (which B. Frénicle had already accomplished). Lord W. Brouncker, an associate of Wallis, apparently solved the problem for $D = 313$. Somehow or other, Euler mistakenly attributed Brouncker's method of solving the equation to J. Pell, who apparently did nothing more than translate or revise someone else's translation of the problem. Refer to Albert H. Beiler, *Recreations in the Theory of Numbers,* for an in-depth discussion.

For additional recreations involving numbers refer to Calendars; Digital Problems; Guessing Numbers.

BIBLIOGRAPHY

Bakst, Aaron. *Mathematical Puzzles and Pastimes.* 2d ed. Princeton, N.J.: D. Van Nostrand Co., 1965.

———. *Mathematics: Its Magic and Mastery.* New York: D. Van Nostrand Co., 1941.

Ball, W. W. Rouse. *Mathematical Recreations and Essays.* 1892. Rev. by H.S.M. Coxeter. London: Macmillan and Co., 1939.

———. *A Short Account of the History of Mathematics.* 1888. Reprint. London: Macmillan and Co., 1935.

Beiler, Albert H. *Recreations in the Theory of Numbers: The Queen of Mathematics Entertains.* 2d ed. New York: Dover, 1966.

Bell, E. T. *The Development of Mathematics.* New York: McGraw-Hill, 1940.

Bowers, Henry, and Joan E. Bowers. *Arithmetical Excursions: An Enrichment of Elementary Mathematics.* New York: Dover, 1961.

Cadwell, J. H. *Topics in Recreational Mathematics.* London: Cambridge University Press, 1970.

Collins, A. Frederick. *Fun with Figures.* New York: D. Appleton and Co., 1928.

Courant, Richard, and Herbert Robbins. *What Is Mathematics? An Elementary Approach to Ideas and Methods.* New York: Oxford University Press, 1941.

Friend, J. Newton. *More Numbers: Fun and Facts.* New York: Charles Scribner's Sons, 1961.

_____. *Still More Numbers: Fun and Facts.* New York: Charles Scribner's Sons, 1964.

Gardner, Martin. *Aha! Gotcha: Paradoxes to Puzzle and Delight.* San Francisco: W. H. Freeman and Co., 1982.

_____. *The "Scientific American" Book of Mathematical Puzzles and Diversions.* New York: Simon & Schuster, 1959.

Kline, Morris. *Mathematics in Western Culture.* New York: Oxford University Press, 1953.

Kordemsky, Boris A. *The Moscow Puzzles.* Trans. by Albert Parry; ed. by Martin Gardner. New York: Charles Scribner's Sons, 1972.

Kraitchik, Maurice. *Mathematical Recreations.* 2d rev. ed. New York: Dover, 1953.

Madachy, Joseph S. *Mathematics on Vacation.* New York: Charles Scribner's Sons, 1966.

Northrop, Eugene P. *Riddles in Mathematics: A Book of Paradoxes.* New York: D. Van Nostrand Co., 1944.

Schaaf, W. L. "Number Games and Other Mathematical Recreations." In *Macropaedia: Encyclopaedia Britannica,* 1985.

Steinhaus, H. *Mathematical Snapshots.* 3d ed. New York: Oxford University Press, 1969.

**NUMEROLOGY,** also called number mysticism, is the study of the mystical or nonscientific meaning of numbers. It might be said to relate to mathematics as astrology relates to astronomy. If all things can ultimately be reduced to numbers, as Pythagorean theory claims, then numerology may, indeed, contain the keys to the universe.

Pythagoras and his school discovered a mathematical relationship between musical pitches, i.e., a numerical law that is true and universal for a basic, natural fact of everyday life. This discovery revealed such a simple underlying logic for a part of nature that Pythagoras jumped to the conclusion that all of nature could be reduced to simple mathematical relationships, including such things as virtue, friendship, love, justice, beauty, truth, and other human values. This is the jump from mathematics to numerology.

The jump has been and continues to be made by some brilliant people. E. T. Bell, *Numerology: The Magic of Numbers,* includes the following translation of a passage from Plato's *Republic,* Book III:

For that which, though created, is divine, a recurring period exists, which is embraced by a perfect number. For that which is human, however, by that one for which it first occurs that the increasings of the dominant and the dominated, when they take three spaces and four boundaries making similar and dissimilar and increasing and decreasing, cause all to appear familiar and expressible.

Whose base, modified, as four to three, and married to five, three times increased, yields two harmonies: one equal multiplied by equal, a hundred times the same: the other equal in length to the former, but oblong, a hundred of the numbers upon expressible diameters of five, each diminished by one, or by two if inexpressible, and a hundred cubes of three. This sum now, a geometrical number, is lord over all these affairs, over better and worse births; and when in ignorance of them, the guardians unite the brides and bridegrooms wrongly, the children will not be well-endowed, either in their constitutions or in their fates.

This "lord" apparently is the number *60 × 60 × 60 × 60* or *12,960,000,* which Plato probably got from the ancient Babylonian priests, who worked with a base 60 mathematical system, instead of a base 10, and who used the number in a similar sense.

Numerological claims are common in Christianity; perhaps the most famous of them is the number of the Beast. St. John, Revelation 13: 18, King James Bible, states:

Here is wisdom. Let him that hath understanding count the number of the beast: for it is the number of a man; and his number is Six hundred threescore and six.

What this means and where the number (666) came from is open to interpretation, and people have been attempting to interpret it for centuries. One common application, called Beasting the Man, has been to assign numbers to the letters in a person's name (or birthday, or whatever) and see if they add up to 666. By careful manipulation it is always possible to Beast someone (dangerous when people truly believe).

Bell includes an account of what is sometimes referred to as the Battle of the Windmills, which took place in the sixteenth century. The two combatants were Michael Stifel, an important German algebraist, and Peter Bung, an encyclopaedist of numerology. Stifel began the battle by attaching 666 to Pope Leo X. Bung replied by attaching the same sign of the Beast to Martin Luther. The battle ended as a standoff. It did not make sense that both the Catholic Church and the anti-Catholic Church should be led by people with the sign of the Beast. Numerology had contradicted itself. And at least for a brief period of time numerology lost some of its glamour.

But the number of the Beast (which continues to reappear in various forms) is only one of dozens of numerological claims derived from the Bible. St. Jerome, for example, claimed that God purposely omitted mention of the second day's creation as "good," because two is an evil number. Philo Judaeus (20 B.C.–A.D. 54) claimed that God had to create all in six days, because six is the most productive of all numbers and, therefore, six is the perfect creative number and cannot be surpassed.

St. Augustine applied the following numerological analysis (a typical example of the type of complex numerology employed by many Biblical numerologists), in this case to emphasize the importance of the number seventeen: The sum of the first seventeen numbers is 153. St. Peter and others once caught 153 fish. If the first seventeen numbers are added, the result is one-half of seventeen times eighteen, or seventeen times nine. Seventeen is ten plus seven. Ten is the law (the Ten Commandments), and seven is the sacred number of the gifts of the spirit. Law without spirit is death; thus, seven must be added to ten. The importance of seventeen is obvious—at least to St. Augustine.

Numerology has also been pursued outside of the Bible. Dr. Wilhelm Fliess (1855–1928) developed a complex system of numerological predicting based on female cycles of 28 days and male cycles of 23 days. By using various combinations of these cycles he determined the best times for all of the activities of a person's life (much the same as astrologers do).

Fliess wrote numerous articles and books on his theory. *Der Ablauf des Lebens: Grundlegung zur Exakten Biologie* (The Rhythm of Life: Foundations of an Exact Biology; 1923) is the most complete expression of his theories. His followers included Sigmund Freud, and by the time of his death a strong cult had been established in Germany. Even published refutations of his system by such highly respected physicists as J. Aelby failed to stem his popularity. The system still had numerous followers as recently as 1970, and undoubtedly still has believers today.

Martin Gardner has published an explanation and refutation of the system in his book *Mathematical Carnival*. Gardner has also created a character, Dr. Matrix, as a spoof on numerology. Dr. Matrix constantly comes up with absurd predictions based on numerology—many of which were first printed in *Scientific American* and later collected in *The Numerology of Dr. Matrix* and *The Incredible Dr. Matrix*.

Some of the common associations that have been attached to the first ten numbers follow:

*One* represents the first principle, God the creator, the east, unity, both male and female, immortality, the right side, the day, the sun, and equality. It is the number from which all other numbers issue; it is the source. It is the basic element, the essence of things.

*Two* represents divisibility, female, mortal, the left side, the night, the moon, inequality, and matter. It is the symbol of opinion. Since it is the lowest divisible prime number and the only even prime number, it is sometimes associated with good luck (though, in general, good luck is associated with odd numbers).

A clover with only two leaflets is a charm that enables its owner to discover his or her future lover by placing it under the sleeper's pillow. The second day of the second month of the year commemorates the Purification of the Virgin Mary in the Temple.

Two may also represent misfortune. People speak of not caring two hoots. One superstition has it that a person will visit each place a second time before he dies (thus, a criminal will revisit the scene of the crime). Some believe that death occurs in twos.

*Three* is regarded as the first odd number and as an extremely positive number. Virgil wrote *"Numero deus impare gaudet"* (God rejoices in the odd number). Shakespeare, *The Merry Wives of Windsor,* wrote: "This is the third time; I hope, good luck lies in odd numbers. . . . They say, there is divinity in odd numbers, either nativity, chance or death."

As the first odd number, three represents the soul. It is associated with the Holy Trinity, the gifts of the Magi, Neptune's trident, and Jove's triple thunderbolt. Cerberus' has three heads, and three brass balls are used to symbolize a pawnshop. People call for three cheers. They bow three times to the new moon. They walk three times around a church altar at midnight to cure fits. They walk through three separate parishes before breakfast to cure the whooping cough. The Church of England demands a threefold publication of a marriage.

Pythagoras put great store in three, because it is the only number that has precisely one beginning, one middle, and one end.

Dante's *Divine Comedy* is divided into three parts, Hell, Purgatory, and Paradise. It has three principle characters, Dante, Virgil, and Beatrice (symbolizing man, reason, and revelation). Each of its three regions is divided into nine (3 × 3) parts.

*Four,* since it is the smallest square (other than one), represents justice and both the body and the soul. For American Indians, it is the sacred number. It has been especially significant for those who believe that all things are made up of four elements—earth, air, fire, and water. It also symbolizes the four cardinal virtues—justice, prudence, temperance, and fortitude.

Also, since it is the smallest square (other than one), it may represent good luck. According to Palestinian folklore, the first four days of every month and the four days following the tenth of every month are especially lucky. A four-leaf clover brings true love.

*Five,* since it is the sum of the first even number (two), which is the first female number, and the first odd number (three), which is the first male number, represented marriage to the Pythagorean school. They also chose a five-pointed star to symbolize their secret mathematical society. Today, the Olympic symbol is made up of five intersecting rings (representing the five continents of the world linked together in harmony).

*Six,* since it is equal to the sum of its divisors (one, two, and three) is called a perfect number. For some it contains the secret of cold.

*Seven* represents completion (e.g., the seven days in a week). It has also been used to represent the most important moments of life, and thus is "lucky" time. It is considered to contain the secret of health.

During the Middle Ages, all of the universe was divided up into sevens: the seven spheres that surrounded the Earth, each containing one of the heavenly bodies (Sun, Moon, and the five known planets), each emitting one of the seven notes of the heavenly music, and so on. Man would pass through these seven spheres when he was born, picking up qualities from each, and then pass through them away from Earth when he died, leaving those qualities behind (disrobing) to appear naked before God upon death.

There are the Seven Wise Men of Greece: Chilo (Know yourself), Cleobulus (Moderation is the greatest good), Pittacus (Know your opportunity), Solon (Nothing in excess), Bias (Too many workers spoil the work), Thales (Suretyship brings ruin), and Periander (Have forethought in all things).

There are the Seven Champions of Christendom (St. George, St. Patrick, St. Andrew, St. David, St. Denys of France, St. Antony of Italy, and St. James of Spain).

There are, on the other hand, the Seven Deadly Sins (pride, lust, covetousness, anger, gluttony, envy, and sloth).

There are the Seven Wonders of the World (the lighthouse of Pharos, Diana's temple at Ephesus, the hanging gardens of Babylon, the pyramids of Egypt, the colossus of Rhodes, the mausoleum at Halicarnassus, and Phidias' statue of Jupiter at Athens).

There are the Seven "Liberal Arts": arithmetic, geometry, astronomy, music, grammar, logic, and rhetoric.

There are the Seven Ages of Man, the Seven Senses, the Seven Seas, and the Seven Heavens. When someone is happy he is said to be in Seventh Heaven.

*Eight* contains the secret of friendship and love. It represented Christ in Cabalist literature. The Greek number for Jesus is 888, the digital root being the human number 6 (thus showing that Jesus contained the human element within himself).

*Nine* was considered a symbol of misfortune by the Pythagorean school, and that belief is also expressed in the nine circles of Dante's Inferno. In *Paradise Lost,* Milton describes Satan and the evil angels lying banished in Hell: "Nine times the space that measures day and night / To mortal men."

Since nine is the first odd square and the first odd number that is not a prime, it may also represent fruitfulness and success. However, this success is often an oversuccess, as in "dressed-up to the nines," meaning overdressed.

*Ten* was used by the Pythagoreans as a symbol of brotherhood. It is the fourth triangular number: $10 = 1 + 2 + 3 + 4$.

One final number deserves mention, and that is thirteen. In America this is almost always considered an unlucky number, possibly because Judas was the thirteenth person at the Last Supper. Thus, all things connected with thirteen are to be avoided: the thirteenth floor, the thirteenth day of the month (especially if it is a Friday), and so on.

In some parts of the world, however, thirteen is a lucky number. In Belgium the number thirteen is worn by women as a good-luck charm. In southern Italy it is also considered a lucky number. The Cabalists associated thirteen with the enemies of God, and yet they considered the thirteenth child in a large family to be a prodigy, a child of singular talent.

For more discussion of numerology associated with letters and words, *see* Gematria. For other mathematical recreations involving the use of numbers to represent letters or words *see* ABC Words; ACE Words; Centurion; Numwords.

BIBLIOGRAPHY

Bell, E. T. *Numerology: The Magic of Numbers.* New York: United Book Guild, 1945.
Bourguignon. Erika. "Numerology." In *Encyclopaedia Americana,* International Ed., 1985.
Bowers, Henry, and Joan E. Bowers. *Arithmetical Excursions: An Enrichment of Elementary Mathematics.* New York: Dover, 1961.
Friend, J. Newton. *More Numbers: Fun and Facts.* New York: Charles Scribner's Sons, 1961.
Frohlichstein, Jack. *Mathematical Fun, Games and Puzzles.* New York: Dover, 1967.
Gardner, Martin. *The Incredible Dr. Matrix.* New York: Charles Scribner's Sons, 1976.
———. *Mathematical Carnival.* New York: Alfred A. Knopf, 1965.
———. *The Numerology of Dr. Matrix.* New York: Simon & Schuster, 1967.

**NUMWORDS** combine mathematics with language.

The letters of the alphabet are sequentially given a corresponding numeral (i.e., A = 1, B = 2, C = 3, and so on). Then either of the following two games may be played.

In the first game, a number is selected (generally one around number 65), and a time limit is established (e.g., two minutes). Each player attempts to come up with words whose numerical value equals the selected number. The player who comes up with the most words wins. The total value of the letters of each word is called the Wordnum. The object, then, of Numwords is to make Wordnums.

The second game is a variation on the first. In this game an established word length (e.g., 5 letters) or an initial letter (e.g., H) is agreed upon. The object is to find the lowest and/or the highest value Numword.

The name Numwords and the two games were thought up by David Parlett and are discussed in his book, *Botticelli and Beyond: Over 100 of the World's Best Word Games.*

For similar mathematical word play refer to ABC Words; ACE Words; Centurion. For mathematical mysticism involving letters and words refer to Numerology.

BIBLIOGRAPHY

Parlett, David. *Botticelli and Beyond: Over 100 of the World's Best Word Games.* New York: Pantheon Books, 1981.

**NYMPHABET.** See NIM.

# O

**OBLONG NUMBERS.** See NUMBER PATTERNS, TRICKS, AND CURIOSITIES.

**ONE LINE NIM.** See NIM.

**OP ART.** See OPTICAL ILLUSIONS.

**OPTICAL ILLUSIONS** involve physiological and psychological considerations as well as mathematical and geometrical principles.

The fluctuations of figure and ground in figures such as the following apparently cannot be avoided once both the vase and the two faces have been recognized:

The Müller-Lyer illusion is based on Gestalt principles of convergence and divergence which make the lines seem to be of different lengths:

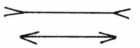

The Necker cube seems to reverse itself:

The parallel lines in the Zöllner illusion seem anything but parallel:

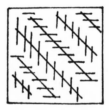

In the Ponzo illusion, the two circles seem to be different sizes:

In the Hering illusion, the horizontal lines really are parallel:

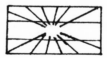

Such optical illusions have brought contemporary artists and mathematics together. Op art is a form of abstract hard-edge art that appeared about the turn of the twentieth century and became the leading form of art in the 1960s. It is based on a strong mathematical order, vivid colors, and straight edges, and results in numerous optical illusions.

Perhaps the best example of "mathematical art" occurs in the lithographs, woodcuts, wood engravings, and mezzotints of Maurits C. Escher of the Netherlands. In *The Graphic Work of M. C. Escher,* p. 8, he writes the following:

The ideas that are basic to [my works] often bear witness to my amazement and wonder at the laws of nature which operate in the world around us. . . . By keenly confronting the enigmas that surround us, and by considering and analyzing the observations that I had made, I ended up in the domain of mathematics.

In *Ascending and Descending* (lithograph, 1960, 38 × 28.5 cm), Escher began with a motif he encountered in an article by L. S. and R. Penrose in the February 1958 issue of the *British Journal of Psychology.* The geometrical inner courtyard is surrounded by a building topped by a never-ending stairway, which monks climb daily, somehow never reaching a top or bottom, but, rather, simply going around and around.

For similar recreations refer to Geometric Dissections; Mirrors.

BIBLIOGRAPHY

Escher, M. C. *The Graphic Work of M. C. Escher: Introduced and Explained by the Artist.* New York: Ballantine Books, 1971.
Gardner, Martin. *Martin Gardner's Sixth Book of Mathematical Games from "Scientific American."* San Francisco: W. H. Freeman and Co., 1963.
_____. *Mathematical Carnival.* New York: Alfred A. Knopf, 1977.
Kingslake, Rudolf, and Brian J. Thompson. "Principles of Optics." In *Macropaedia: Encyclopaedia Britannica,* 1985.
Schaaf, W. L. "Number Games and Other Mathematical Recreations." In *Macropaedia: Encyclopaedia Britannica,* 1985.
West, Louis J. "Perception." In *Macropaedia: Encyclopaedia Britannica,* 1985.

**ORIGAMI.** See GEOMETRIC PROBLEMS AND PUZZLES.

**OXYMORON.** See LOGICAL PARADOXES.

# P

**PANMAGIC SQUARES.** See MAGIC SQUARES.

**PAPER FOLDING PUZZLES.** See GEOMETRIC PROBLEMS AND PUZZLES.

**PARADOX.** See LOGICAL PARADOXES.

**PARADOXES OF THE INFINITE** are problems, puzzles, and amusements based on various attempts to explain infinity (limitlessness) in time and space.

Morris Kline (*Mathematics in Western Culture,* p. 395), includes the following quotation from Voltaire:

We admit, in geometry, not only infinite magnitudes, that is to say, magnitudes greater than any assignable magnitude, but infinite magnitudes infinitely greater, the one than the other. This astonishes our dimension of brains, which is only about six inches long, five broad, and six in depth, in the largest brain.

Galileo stated that it is impossible to compare infinite quantities. After all, he said, the number of whole numbers is infinite (larger than any finite number that can be named), and the number of even whole numbers is also infinite. Since the number of whole numbers contains all of the numbers of the number of whole even numbers, the number of whole numbers should be the larger set. However, for each number in one set there corresponds one number in the second set (a one-to-one correspondence). Thus, there should be as many in one set as in the other. This, then, has resulted in a paradox (a statement or proposition that appears to be self-contradictory).

According to W. L. Schaaf, "Number Games and Other Mathematical Recreations," a "mathematical paradox is a mathematical conclusion so unexpected that it is difficult to accept even though every step in the reasoning is valid." It is distinguished from a "mathematical fallacy," which he defines as "an instance of improper reasoning leading to an unexpected result that is patently false or absurd," and includes a "mathematical sophism," defined as "a fallacy in which the error has been knowingly committed." The most famous sophisms involving infinity of space and time are those of Zeno (*see* Zeno's Paradoxes).

David Hilbert put forth the following paradox of infinite space, titled "Hotel Infinity" by Martin Gardner and discussed in his book, *Aha! Gotcha: Paradoxes to Puzzle and Delight*: Hotel Infinity is located at the center of the galaxy. It has an infinite number of rooms that extend through a black hole to a higher dimension. The rooms start at number one.

One day, when all of the rooms have been filled, a space traveler arrives and asks for a room. "No problem," says the hotel manager, "I'll simply move each occupant to a room with one higher number and give you room number one."

Later that day, six Martians arrive, looking for separate rooms. "No problem," says the hotel manager, "I'll just move each occupant to a room with a number six higher than the room he currently occupies and give you the first six rooms.

One weekend, however, an *infinite* number of conventioneers arrive from Pluto, each wanting a separate room. Can the hotel manager solve this problem? "Of course," says the hotel manager. "I'll simply move everyone to a room with a number twice as large as before, thus putting everyone in a room with an even number and leaving all the odd numbered rooms empty."

Therefore, infinity subtracted from infinity still leaves infinity.

Morris Kline includes the following paradox on infinity of time: Tristram Shandy was hopelessly perplexed. He had begun to write his autobiography and found that he could record only half a day's experiences in one day of writing. Consequently, even if he were to start writing at birth and even if he were to live forever, he could not record his whole life, for at any time only half of his life would be recorded. And yet if he did live on indefinitely he ought to be able to record his whole life, for the experiences of his first ten years would be recorded by the end of his twentieth year; the experiences of his first twenty years by the end of his fortieth year; and so on. Thereby every year of his life would be reached at some time. Hence, depending on which way he reasoned, he could or could not complete his autobiography. The longer Tristram puzzled over this paradox the more confused he became and the farther he appeared to be from a decision.

Whereas many of the problems raised by the ancient Greek mathematicians (e.g., Doubling the Cube, Euclid's Theory of Parallels, Squaring the Circle, and the Trisection of an Angle; refer to each of these entries for further discussion) have been resolved to the satisfaction of most contemporary mathematicians, the paradoxes of infinity are still unresolved to universal agreement.

Nevertheless, progress has been made, and the theories of Georg Ferdinand Ludwig Philipp Cantor (1873) have gained strong support, if not as a complete solution, at least as a more useful approach. Cantor realized that counting the number of items in an infinite collection simply results in an endless task. Moreover, he recognized the deeper significance of one-to-one correspondence (i.e., it is not necessary to count the number of items in a collection to establish that collections with a one-to-one correspondence must have the same number of items). Therefore, according to Cantor, if two infinite collections (classes) can be put into a one-to-one correspondence, then they have the same number of objects in them. Cantor then designated the number of items in the infinite collection of all positive integers by $\aleph_0$ (aleph-null). This is the smallest infinite cardinal number. This, then, according to Cantor, solves Galileo's problem (i.e., the number of whole numbers and the number of whole even numbers is the same because the two collections of numbers are in one-to-one correspondence).

This seeming contradiction, according to Cantor, is only so when dealing with finite collections. On the contrary, when dealing with infinite numbers it is the rule. In fact, the definition of infinite collections is that they can be put into a one-to-one correspondence with a part of themselves, which is not the case with finite numbers.

Cantorism (also called the Continuum Hypothesis) is that there is an infinite number of higher infinities beyond aleph-null. The explanation goes as follows:

Any infinite set of things that is countable (e.g., 1, 2, 3 . . .) is an aleph-null set. Each subset of a set of things (e.g., the number 3) contains a subset equal to 2 to the power of the number of things in the set (e.g., the subset of 3 contains subsets equal to $2^3$ or 8). Thus, an aleph-null set contains subsets equal to 2 to the power of aleph-null.

For example: A finite set of three things, say, the jack, queen, and king of hearts, produces the following eight subsets (the original set, the empty set, three variations of the one-card set, and three variations of the two-card set; $x$ means the card is in the set; $o$ means it is not in the set):

| Jack | Queen | King |
| --- | --- | --- |
| x | x | x |
| o | o | x |
| o | x | o |
| x | o | o |
| x | x | o |
| o | x | x |
| x | o | x |
| o | o | o |

A set of *1* (e.g., if the set consisted only of the king of hearts) contains $2^1$ or *2* subsets (one with the king in it and one without the king in it). A set of *2* contains $2^2$ or *4* subsets. A set of *3* contains $2^3$ or *8* subsets. And so on.

This process can be extended to show that the set of real numbers (numbers that can be expressed exactly by the ratio of two integers [i.e., rational numbers], or as the limit of a sequence of rational numbers [i.e., irrational numbers] as opposed to complex numbers [i.e., numbers expressed as *(a + bi)* in which *a* and *b* are real numbers and *i* is defined as $\sqrt{-1}$]) is uncountable, i.e., aleph-one.

If the cards are replaced by binary numbers (*1* if in the set, *0* if not), then by placing a decimal point in front of each row (extending the cards to an aleph-null set) it is possible to obtain an infinite number of binary fractions between *0* (an empty set) and *1* (a full set). By reversing the *0*'s and *1*'s, a new list of binary fractions not on the original list is obtained.

Furthermore, if these points are placed on a line, another set of nonrational points (e.g., $\pi$ and $\sqrt{2}$) can also be included. Therefore, this line contains three sets in one-to-one correspondence: the subsets of aleph-null, the set of real numbers, and the total number of points on the line. Cantor gave this higher infinity the cardinal number *C* (the power of the continuum) and stated that it was aleph-one (the first infinity greater than aleph-null).

Cantor went on to show that *C* is the number of such infinite sets as transcendental irrationals (algebraic irrationals form a countable set, i.e., are aleph-null), the number of points on a line of infinite length, the number of points on any plane figure or an infinite plane, and the number of points on any solid figure or in all of three-space.

Morris Kline, *Mathematics in Western Culture,* includes an explanation of Cantor's proof of *C* for an infinite line. It goes as follows:

*C* equals the number of numbers between *0* and *1*. Thus, any collection of items in a one-to-one correspondence with all the numbers between *0* and *1* must also contain *C* items.

The points on a line segment are used as an example of a collection of *C* items. A fixed point *0* is established on a line and each point on the line is designated by a number that expresses the distance of that point from *0*, distances to the right expressed in positive numbers, those to the left in negative numbers. *C* equals the number of real numbers between *0* and *1*. There is a one-to-one correspondence between the numbers between *0* and *1* and the points on the line which those numbers designate.

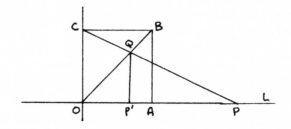

*C* is in a one-to-one correspondence with all positive real numbers. The set of real numbers is in one-to-one correspondence with the points on a line. Let the points on line *L* to the right of *0* represent all the positive real numbers, and let *0A* be the unit segment such that its points are in one-to-one correspondence with the real numbers between *0* and *1*. Construct rectangle *0ABC* and draw diagonal *0B*. *P* equals any point to the right of *0*. Draw *CP* such that it intersects *0B* at *Q*. Draw a perpendicular from *Q* to intersect *0P* at *P'*. Under the correspondence established, any point *P* anywhere on *L* to the right of *0* corresponds to one and only one point *P'* on *0A*. Conversely, any point *P'* on *0A* will form a perpendicular that will intersect *0B* at *Q*. Draw *CQ*. Where *CQ* intersects *L* point *P* is obtained corresponding to *P'*. Since the points on *0A* are in one-to-one correspondence with all of the points to the right of *0* on *L*, the number of points on *0A* as well as the entire half-line equals *C*. Thus, the collection of positive real numbers is in one-to-one correspondence with the real numbers between *0* and *1*. Thus, the number of positive real numbers is *C*. In other words, the number of points on a line of infinite length is equal to the number of points on a line segment.

This concept that any two line segments, regardless of their lengths, possess the same number of points, allows a solution of Zeno's paradoxes (*see* Zeno's Paradoxes). For example, Zeno's paradox (or sophism) of Achilles and the Tortoise states that Achilles and a tortoise run a race. The tortoise is given a head start because it is so much slower than Achilles. The race will end when Achilles overtakes the tortoise. At each instant in the race Achilles and the tortoise are at some point in their paths. Since each runs for the same number of instants, the tortoise runs through as many distinct points as Achilles. Achilles, however, must run through more points than the tortoise. Therefore, he can never overtake the tortoise. However, according to Cantor's theories, Achilles does not have to run through more points than the tortoise, i.e., distance has nothing to do with the number of points.

Tristram Shandy's problem can also be solved. Since he will live 2 × aleph-null years, he can record aleph-null years of his life. Applying Cantor's theories, 2 × aleph-null = aleph-null, so the task can be accomplished.

The paradoxes of the infinite have been solved. Or have they? Consider the following: The collection of all positive integers equals aleph-null. The collection of all positive integers after number 11 equals aleph-null. Thus, aleph-null − 11 = aleph-null.

Is the Continuum Hypothesis correct? David Hilbert, a great twentieth-century mathematician, has stated, "No one shall expel us from the paradise which Cantor has created for us" (Morris Kline, *Mathematics in Western Culture*, p. 397). On the other hand, Henri Poincaré, another great twentieth-century mathematician, has stated, "Later generations will regard [Cantor's] *Mengenlehre* as a disease from which one has recovered" (Kline, *Mathematics in Western Culture*, p. 397).

Richard Schlegel, a physicist, has pointed out a contradiction in the theory. According to the "steady-state" theory of the universe, the number of atoms in the cosmos at present is aleph-null (assuming that the cosmos is infinite). Also, the atoms are increasing as the universe expands. Now, if the increasing continues for an aleph-null number of times, the result is 2 raised to the power of aleph-null aleph-null number of times (an aleph-one set). However, any collection of distinct physical entities is countable and therefore, at most, aleph-null. Thus, there is a contradiction.

Schlegel offers a solution. Consider the past as still uncompleted, as is the future. Thus, moving backward in time by halving the atoms means that the atoms can never be halved more than a "finite" number of times (never less than aleph-null), and the same holds for doubling when moving into the future. Thus, we remain at aleph-null.

In 1938 Kurt Gödel proved that Cantor's assumption that there is no aleph between aleph-null and $C$ could be assumed to be true and would not conflict with set theory. In 1963 Paul J. Cohen proved that the opposite could also be assumed, i.e., that there is at least one aleph between aleph-null and $C$. Both Gödel and Cohen believe that ultimately the Continuum Hypothesis will be proven false. However, at the moment the two possibilities exist side by side, dividing the theory of infinite sets into Cantorian and non-Cantorian.

For similar activities refer to Fallacies; Logical Paradoxes; Logical Problems and Puzzles; Zeno's Paradoxes.

BIBLIOGRAPHY

Bakst, Aaron. *Mathematics: Its Magic and Mastery.* New York: D. Van Nostrand Co., 1941.
Collins, A. Frederick. *Fun with Figures.* New York: D. Appleton and Co., 1928.
Gardner, Martin. *Aha! Gotcha: Paradoxes to Puzzle and Delight.* San Francisco: W. H. Freeman and Co., 1982.
_____. *Martin Gardner's Sixth Book of Mathematical Games from "Scientific American."* San Francisco: W. H. Freeman and Co., 1971.
_____. *Mathematical Carnival.* New York: Alfred A. Knopf, 1965.
_____. *Wheels, Life and Other Mathematical Amusements.* San Francisco: W. H. Freeman and Co., 1983.
Grunbaum, Adolf. *Modern Science and Zeno's Paradoxes.* Middletown, Conn.: Wesleyan University Press, 1967.
Kline, Morris. *Mathematics in Western Culture.* New York: Oxford University Press, 1953.
Northrop, Eugene P. *Riddles in Mathematics: A Book of Paradoxes.* New York: D. Van Nostrand Co., 1944.
Russell, Bertrand. *Wisdom of the West.* London: Crescent Books, 1977.
Salmon, Wesley C. *Zeno's Paradoxes.* New York: Bobbs-Merrill Co., 1970.
Schaaf, W. L. "Number Games and Other Mathematical Recreations." In *Macropaedia: Encyclopaedia Britannica,* 1985.

Tietze, Heinrich. *Famous Problems of Mathematics: Solved and Unsolved Mathematical Problems from Antiquity to Modern Times*. Trans. by Beatrice Kevitt Hofstadter and Horace Komm. New York: Graylock Press, 1965.

**PARADOXES OF ZENO.** See ZENO'S PARADOXES.

**PARADOX OF TRISTRAM SHANDY.** See PARADOXES OF THE INFINITE.

**PASCAL'S TRIANGLE.** See NUMBER PATTERNS, TRICKS, AND CURIOSITIES.

**PEG SOLITAIRE.** See COIN PROBLEMS AND PUZZLES.

**PELLIAN EQUATION.** See NUMBER PATTERNS, TRICKS, AND CURIOSITIES.

**PENTAGONAL NUMBERS.** See NUMBER PATTERNS, TRICKS, AND CURIOSITIES.

**PENTOMINO PROBLEMS.** See DOMINOES.

**PERFECT DIGITAL INVARANTS.** See NUMBER PATTERNS, TRICKS, AND CURIOSITIES.

**PERFECT NUMBERS.** See NUMBER PATTERNS, TRICKS, AND CURIOSITIES.

**PERPETUAL CALENDAR.** See CALENDARS.

**PIGPEN CIPHER.** See CRYPTARITHMS, CRYPTOGRAPHY, CONCEAL-MENT CIPHERS, SUBSTITUTION CIPHERS, TRANSPOSITION CIPHERS, AND CODE MACHINES.

**PLAYFAIR CIPHER.** See CRYPTARITHMS, CRYPTOGRAPHY, CON-CEALMENT CIPHERS, SUBSTITUTION CIPHERS, AND CODE MACHINES.

**POLYALPHABETIC CIPHER.** See CRYPTARITHMS, CRYPOTGRAPHY, CONCEALMENT CIPHERS, SUBSTITUTION CIPHERS, AND CODE MACHINES.

**POLYBIUS CHECKERBOARD CIPHER.** See CRYPTARITHMS, CRYPTOGRAPHY, CONCEALMENT CIPHERS, SUBSTITUTION CIPHERS, TRANSPOSITION CIPHERS, AND CODE MACHINES.

**POLYGONAL NUMBER SERIES.** See NUMBER PATTERNS, TRICKS, AND CURIOSITIES.

**POLYIMONOES.** See DOMINOES.

**PORTA'S DIGRAPHIC CIPHER.** See CRYPTARITHMS, CRYPTOGRAPHY, CONCEALMENT CIPHERS, SUBSTITUTION CIPHERS, TRANSPOSITION CIPHERS, AND CODE MACHINES.

**POSTULATE OF PARALLELS.** See EUCLID'S THEORY OF PARALLELS.

**PRIME NUMBERS.** See NUMBER PATTERNS, TRICKS, AND CURIOSITIES.

**PROBABILITY PUZZLES.** See ARITHMETIC AND ALGEBRAIC PROBLEMS AND PUZZLES.

**PROBLEM OF THE FOUR *N*'s.** See DIGITAL PROBLEMS.

**PROGRESSIVE CHESS.** See FAIRY CHESS.

**PROPER FRACTIONS.** See NUMBER PATTERNS, TRICKS, AND CURIOSITIES.

**PURE MATHEMATICS.** See LOGICAL PROBLEMS AND PUZZLES.

**PYRAMIDAL NUMBERS.** See NUMBER PATTERNS, TRICKS, AND CURIOSITIES.

**PYTHAGOREAN THEOREM.** See GEOMETRIC DISSECTIONS.

**PYTHAGOREAN TRIPLES.** See NUMBER PATTERNS, TRICKS, AND CURIOSITIES.

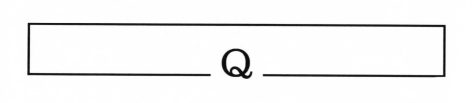

**QUADRATURE OF THE CIRCLE.** See SQUARING THE CIRCLE.

# R

**THE RACECOURSE.** See ZENO'S PARADOXES.

**RAIL FENCE CIPHER.** See CRYPTARITHMS, CRYPTOGRAPHY, CON-CEALMENT CIPHERS, SUBSTITUTION CIPHERS, TRANSPOSITION CIPHERS, AND CODE MACHINES.

**RANDOM SUBSTITUTION CIPHER.** See CRYPTARITHMS, CRYP-TOGRAPHY, CONCEALMENT CIPHERS, SUBSTITUTION CIPHERS, TRANSPOSITION CIPHERS, AND CODE MACHINES.

**REBUS.** See CRYPTARITHMS, CRYPTOGRAPHY, CONCEALMENT CIPHERS, SUBSTITUTION CIPHERS, TRANSPOSITION CIPHERS, AND CODE MACHINES.

**RECURRING DIGITAL INVARIANT.** See NUMBER PATTERNS, TRICKS, AND CURIOSITIES.

**ROMAN REPUBLIC CALENDAR.** See CALENDARS.

**ROUND-THE-CLOCK.** See DOMINOES.

**ROYAL GAME.** See CHESS.

**RUNNERS.** See ZENO'S PARADOXES.

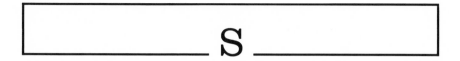

# S

**SCOTCH CHESS.** See FAIRY CHESS.

**SCRAMBLING WITH A KEY WORD CIPHER.** See CRYPTARITHMS, CRYPTOGRAPHY, CONCEALMENT CIPHERS, SUBSTITUTION CIPHERS, TRANSPOSITION CIPHERS, AND CODE MACHINES.

**SCYTALE.** See CRYPTARITHMS, CRYPTOGRAPHY, CONCEALMENT CIPHERS, SUBSTITUTION CIPHERS, TRANSPOSITION CIPHERS, AND CODE MACHINES.

**SEBASTOPOL.** See DOMINOES.

**SELFMATE.** See FAIRY CHESS.

**SEQUENCE OR SERIES PROBLEMS AND PUZZLES.** See ARITHMETIC AND ALGEBRAIC PROBLEMS AND PUZZLES.

**THE SHADOW'S CODE.** See CRYPTARTHMS, CRYPTOGRAPHY, CONCEALMENT CIPHERS, SUBSTITUTION CIPHERS, TRANSPOSITION CIPHERS, AND CODE MACHINES.

**SHIFT CIPHER.** See CRYPTARTHMS, CRYPTOGRAPHY, CONCEALMENT CIPHERS, SUBSTITUTION CIPHERS, TRANSPOSITION CIPHERS, AND CODE MACHINES.

**SNIFF.** See DOMINOES.

**SOMA CUBES.** See TANGRAMS.

**SOPHISMS OF ZENO.** See ZENO'S PARADOXES.

**SPACE CHESS.** See FAIRY CHESS.

**SPACE PUZZLES.** See GEOMETRIC PROBLEMS AND PUZZLES.

**SPEED AND DISTANCE PUZZLES.** See ARITHMETIC AND AL-GEBRAIC PROBLEMS AND PUZZLES.

**SPROUTS.** See TOPOLOGY.

**SQUARED RECTANGLES.** See GEOMETRIC DISSECTIONS.

**SQUARE NUMBERS.** See NUMBER PATTERNS, TRICKS, AND CURIOSITIES.

**SQUARING THE CIRCLE,** also called quadrature of the circle, challenges the mathematician to construct a square containing the same area as that enclosed by a circle. It is equivalent to the problem of rectification of the circle, i.e., of determining a straight line whose length is equal to that of the circumference of a circle. It stands along with the trisection of an angle and the duplication of a cube as one of the three great construction problems of ancient geometry.

The problem is as follows. Find a method of enclosing a square in the same area as that of a given circle using nothing more than a straight-edge and compasses to construct a finite number of circles and lines. In analytical geometry the problem is to find an algebraic expression which will allow for the construction by ruler and compasses of a square whose area equals the area $\pi R^2$ of a circle of given radius $R$. Pi ($\pi$) equals the ratio of the circumference of a circle to its diameter (3.141592 . . .). If the solution is possible, and $R$ is taken as the unit of length, the number $\pi$ of the units of area will satisfy an algebraic equation with rational coefficients.

E. W. Hobson, "Squaring the Circle," divides a historical survey of the problem into three periods distinguished from each other by "fundamentally distinct differences in respect of method, of immediate aims, and of equipment in the possession of intellectual tools" (p. 10). The first period extends from the earliest attempts at an empirical determination of $\pi$ (approximately 1700 B.C.) until Isaac Newton (1642–1727) and Gottfried Wilhelm Leibnitz (or Leibniz) (1646–1716) founded differential and integral calculus.

The earliest attempts to determine $\pi$ on record are in the papyrus *Rhind*, copied by a clerk named Ahmes of the king Raaus, about 1700 B.C., which compares the area of a circle to that of a square whose side is diminished by one-ninth—probably an empirical determination. It was translated and explained by Eisenlohr, *Ein mathematisches Handbuch der alten Agypter* (Leipzig, 1877), and summarized in Hobson's "Squaring the Circle."

Hobson also discusses the less accurate assumption held by the Babylonians ($\pi = 3$), which he suggests was probably connected to their realization that a regular hexagon inscribed in a circle has its side equal to the radius, and a division of the circumference into $6 \times 60 = 360$ equal parts. Hobson points out that the same assumption was implied in the Old Testament, 1 Kings 7:23, and in Chronicles 4:2: "Also he made a molten sea of ten cubits from brim to brom, round in compass, and five cubits the height thereof; and a line of thirty cubits did compass it round about." He further states that the Talmud makes a similar assumption: "That which in circumference is three hands broad is one hand broad."

The Greek mathematicians were the first to give the problem a systematic treatment. Plutarch, *De exilio,* credits Anaxagoras of Clazomene (500–428 B.C.) with the construction of the problem while incarcerated in prison. Hippias of Elis came up with a curve, the quadratrix, which, if it could be constructed, would give a construction for the length $2a/\pi$, and hence for $\pi$. However, the construction of the curve involves the same problem as that of $\pi$. The Sophists attempted to solve the problem by connecting it with "cyclical square numbers" (i.e., $25 = 5^2$).

Bryson and Antiphon, contemporaries of Socrates (469–399 B.C.) came up with close approximations, Antiphon through the inscription of polygons with ever greater numbers of sides, until the inscribed polygon was so close to a circle that "for practical purposes" there was no difference (and it had already been proven by the school of Pythagoras that a polygon could be converted to a square of equal area); Bryson of Heraklea by considering the area of a circle to be the arithmetical mean between an inscribed and circumscribed polygon of equal numbers of sides, thus introducing the idea of minimum and maximum limits to approximations. According to Hermann Schubert, "The Squaring of the Circle," Bryson's comparison of the circle with inscribed and circumscribed polygons "indicated to Archimedes the way by which an approximate value for $\pi$ was to be reached."

In the second half of the fifth century B.C., Hippocrates of Chios, who lived in Athens, was able to show that certain curvilinearly bounded plane figures (the menisci or lunulae of Hippocrates) would allow exact quadrature. However, he was unable to convert a circle into this crescent-shaped figure.

Archimedes, born in Syracuse in 287 B.C. and killed by a Roman soldier "whom he had forbidden to destroy the figures he had drawn in the sand" in 212 B.C. (Schubert), gave the first successful computation of $\pi$, "less than $3\frac{1}{7}$ but greater than $3\frac{10}{71}$" (Archimedes, "Measurement of a Circle: Proposition 3

[Approximation of Using in Essence Upper and Lower Limits]'' translated and reprinted in Ronald Calinger, ed., *Classics of Mathematics*). Archimedes arrived at this close approximation by determining the perimeters of inscribed and circumscribed polygons of ever greater numbers of sides, employing the ''method of exhaustions'' which is stated in Euclid x. 1:

Two unequal magnitudes being set out, if from the greater there be subtracted a magnitude greater than its half, and from that which is left a magnitude greater than its half, and if this process be repeated continually, there will be left some magnitude which will be less than the lesser magnitude set out.

The rest of this ''first period'' is little more than a list of mathematicians who came up with ever closer approximations of squaring the circle without making any real advancements on Archimedes' theories. Ptolemy (87–165) developed a trigonometry that allowed a more exact approximation to $\pi$. Leonardo Pisano (thirteenth century) reached limits of $\dfrac{1440}{458\frac{1}{8}}$ and $\dfrac{1440}{458\frac{4}{9}}$. Adriaen Anthonisz (1527–1607) rediscovered the Chinese value 3.1415929 (correct to six decimal places). François Viète (1540–1603) calculated $\pi$ to nine decimal places, Adrianus Romanus (1561–1615) to fifteen, Ludolf van Ceulen (1539–1610) to thirty-five.

The second period, beginning in the second half of the seventeenth century, turned from geometrical or semi-geometrical methods of determining $\pi$ to analytical expressions formed according to definite laws. John Wallis (1616–1703), *Arithmetica Infinitorum*, came up with $\dfrac{\pi}{8}$ as the area of a semicircle of diameter 1 equalling the definite integral $\int_0^1 \sqrt{x-x^2}dx$. From this he came up with:

$$\frac{\pi}{2} = \frac{2}{1} \cdot \frac{2}{3} \cdot \frac{4}{3} \cdot \frac{4}{5} \cdot \frac{6}{5} \cdot \frac{6}{7} \cdot \frac{8}{7} \cdot \frac{8}{9} \cdots$$

for $\pi$ as an infinite product. Furthermore, he showed that the approximation obtained by stopping at any fraction is less than or greater than $\dfrac{\pi}{2}$ in accordance with whether the fraction is proper or improper.

Though Newton and James Gregory had already discovered it, in 1624 Leibnitz came up with what has since been called Leibnitz's series, $\dfrac{\pi}{4} = 1 - \dfrac{1}{3} + \dfrac{1}{5} - \ldots$, and published it in ''De vera proportione circuli ad quadratum circumscriptum in numeris rationalibus'' (1682), indicating how it related to a representation of $\pi$.

By substituting the values $\dfrac{\pi}{6}, \dfrac{\pi}{8}, \dfrac{\pi}{10}, \dfrac{\pi}{12}$ in Gregory's more general series,

$$a = t - \frac{t^3}{3r^2} + \frac{t^5}{5r^2} - \ldots \quad (a = \text{ the length of an arc}, t = \text{ the length of a tangent}$$

at one extremity of the arc, and $r =$ the radius of the circle), Abraham Sharp was able to calculate $\pi$ up to seventy-two places.

Newton's series was based on the following: $\sin^{-1}x = x + \frac{1}{2} \times \frac{x^3}{3} + \frac{1 \cdot 3}{2.4} \times \frac{x^5}{5} + \cdots$

Owing to the coefficients, however, it was a more difficult one to work with.

In 1737, Leonhard Euler connected up trigonometrical ratios with the exponential function, coming up with the relationship between $e$ and $\pi$ ($e^{i\pi} = -1$) that would later lead to establishing the true nature of $\pi$.

The third period began in the middle of the eighteenth century and lasted until late in the nineteenth century. During this period, mathematicians began studying the nature of $\pi$ itself, first establishing that it is irrational and then that it is transcendental, thus proving that it *cannot* be constructed by Euclidean determination.

J. H. Lambert (1728–1777) proved that $\pi$ is an irrational number, establishing the following theorems:

1. If $x$ is a rational number, different from zero, $e^x$ cannot be a rational number.

2. If $x$ is a rational number, different from zero, tan $x$ cannot be a rational number.

In 1840, Joseph Liouville proved the existence of transcendental numbers (numbers which can be defined which cannot be the root of any algebraical equation with rational coefficients). G. Cantor (*Crelle's Journal*, vol. 77, 1874) proved the same by showing that there are numbers which do not belong to the sequence of algebraic numbers and are therefore transcendental.

Analytical geometry replaces every geometrical problem with a corresponding analytical one involving only numbers and their relations. In Euclidean geometry, the determination of a point is accomplished by a finite number of applications of three processes: (1) the intersection of straight lines each given by a pair of already determined points; (2) the intersection of a straight line given by two points and a circle given by its center and one point on the circumference, all four points having already been determined; and (3) the intersection of two circles which are determined by four points already determined. In analytical geometry, an original set of numbers is given $(a_1, a_2, \ldots a_{2r})$, and the coordinates of $r$ given points; $(r \geq 2)$. Then, at each stage of the geometrical process two new numbers are determined, the coordinates of a new point.

Since the determination of point $P$ is to be made by a finite number of steps, the coordinates of $P$ are determined by means of a finite succession of operations on $(a_1, a_2, \ldots a_{2r})$, the coordinates of the points, and each of these operations consists either of a rational operation or of one involving the process of taking a square root of a rational function as well as a rational operation. Thus, if $P$ can be determined by Euclidean geometry it is necessary and sufficient that its coordinates can be expressed as functions of the coordinates $(a_1, a_2, \ldots a_{2r})$ of the given points of the problem as will involve in successive performance a finite number of times operations which are either rational or involve taking a square root of a rational function of elements already determined.

Furthermore, it has been shown that for a point $P$ to be determined by Euclidean procedure it is necessary that each of its coordinates be a root of an equation of some degree, a power of 2, of which the coefficients are rational functions of $(a_1, a_2, \ldots a_{2r})$, the coordinates of the points given in the data of the problem.

In the case of squaring the circle, the given consists of two points $(0,0)$ and $(1,0)$, and the point to be determined has coordinates $(\pi,0)$. Since $\pi$ is not a root of any algebraic equation, it is not able to be arrived at through Euclidean construction.

However, squaring the circle remains one of the major pastimes of amateur mathematicians who refuse to believe the findings of analytical geometry. As Schubert states, "The race of circle-squarers, unmindful of the verdict of mathematics, that most infallible of arbiters, will never die out so long as ignorance and the thirst for glory shall be united."

For similar mathematical recreations refer to The Duplication of the Cube; Geometric Problems and Puzzles; The Trisection of an Angle.

BIBLIOGRAPHY

Ball, W. W. Rouse. *Mathematical Recreations and Essays*. 1892. Rev. by H.S.M. Coxeter. London: Macmillan and Co., 1939.

————. *A Short Account of the History of Mathematics*. London: Macmillan and Co., 1888.

Bell, E. T. *The Development of Mathematics*. New York: McGraw-Hill, 1945.

Cadwell, J. H. *Topics in Recreational Mathematics*. London: Cambridge University Press, 1966.

Calinger, Ronald, ed. *Classics of Mathematics*. Oak Park, Ill.: Moore Publishing Co., 1982.

Carslaw, H. S. "On the Constructions Which Are Possible by Euclid's Methods." *Mathematical Gazette* 5 (1910): 171–179.

Clark, M. E. "Construction with Limited Means." *American Mathematical Monthly* 48 (1941): 475–479.

Dickson, L. E. "Constructions with Ruler and Compasses: Regular Polygons." *Monographs on Topics of Modern Mathematics Relevant to the Elementary Field*. New York: Dover, 1955.

Gow, James. *A Short History of Greek Mathematics*. London: Cambridge University Press, 1884.

Heath, Thomas, trans. *The Thirteen Books of Euclid's Elements Translated from the Text of Heiberg with Introduction and Commentary*. Vol. I. New York: Dover, 1956.

Hobson, E. W. "Squaring the Circle: A History of the Problem." *Squaring the Circle and Other Monographs*. New York: Chelsea Publishing Co., 1969.

Kline, Morris. *Mathematics in Western Culture*. New York: Oxford University Press, 1953.

Schubert, Hermann. "The Squaring of the Circle: An Historical Sketch of the Problem from the Earliest Times to the Present Day." *The Monist* 1 (1891): 197–228.

**SQUARING THE SQUARE.** See GEOMETRIC DISSECTIONS.

**THE STADIUM.** See ZENO'S PARADOXES.

**STOMACHION.** See TANGRAMS.

**STRIP DISSECTION.** See GEOMETRIC DISSECTIONS.

**STROBOGRAMMATIC NUMBERS.** See NUMBER PATTERNS, TRICKS, AND CURIOSITIES.

**SUBSTITUTION CIPHERS.** See CRYPTARITHMS, CRYPTOGRAPHY, CONCEALMENT CIPHERS, TRANSPOSITION CIPHERS, AND CODE MACHINES.

**SUCCOTASH.** See TOPOLOGY.

**SUPERTASKS.** See ZENO'S PARADOXES.

**SYMPATHETIC NUMBERS.** See NUMBER PATTERNS, TRICKS, AND CURIOSITIES.

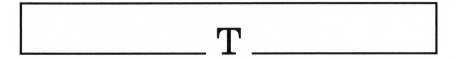

# T

**TAC TIX.** See NIM.

**TANGRAMS** are an ancient Oriental amusement that involve rearranging of a set of geometric shapes into various objects.

The geometric shapes (called *tans*) are cut from a square as follows:

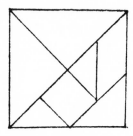

They are then arranged in the shape of a person, an animal, or whatever:

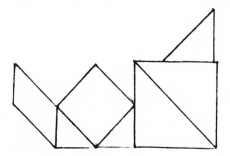

I call the above a geometric mouse. Obviously, tangrams require a bit of "stretching" from the viewer.

The Loculus of Archimedes or Stomachion is a similar set of geometric shapes (double the number) cut from a rectangle whose length is twice that of its width:

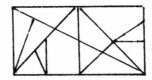

Piet Hein has employed the rules of tangrams to create similar topological play in three dimensions. According to Martin Gardner, *The Second "Scientific American" Book of Mathematical Puzzles and Diversions,* Hein conceived of the *Soma Cube* while listening to Werner Heisenberg lecture on quantum physics. Hein came up with the following question. Can all of the irregular shapes formed by combining no more than four cubes, all of the same size and joined at their faces, be so put together that they form one larger cube?

Since an irregular shape must have a concavity or a corner nook in it somewhere, there are only six different arrangements of irregular shapes possible with four cubes face to face:

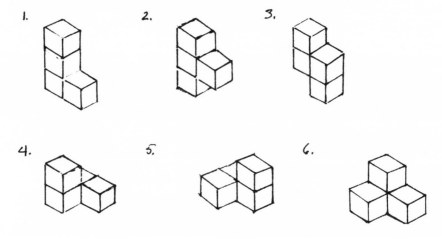

As Hein has indicated, two cubes can be joined at only a single coordinate, three cubes can add a second coordinate perpendicular to the first, and four cubes are necessary to supply the third coordinate perpendicular to the other two. All of the pieces can then be combined to form a cube. According to Gardner, Richard K. Guy of the University of Malaya, Singapore, has found over 230 different solutions (though the total number of possible solutions has not yet been found).

Soma Cubes have been marketed by the Gem Color Company, 200 Fifth Avenue, N.Y., through the authorization of Hein.

For similar play refer to Geometric Dissections; Graphs and Networks; Mazes and Labyrinths; The Möbius Strip.

BIBLIOGRAPHY

Gardner, Martin. *The Second "Scientific American" Book of Mathematical Puzzles and Diversions.* New York: Simon & Schuster, 1961.
Schaaf, W. L. "Number Games and Other Mathematical Recreations." In *Macropaedia: Encyclopaedia Britannica,* 1985.

**'T' CARD PUZZLE.** See CARD TRICKS AND PUZZLES.

**TELEPHONE DIAL CODES.** See CRYPTARITHMS, CRYPTOGRAPHY, CONCEALMENT CIPHERS, SUBSTITUTION CIPHERS, TRANSPOSITION CIPHERS, AND CODE MACHINES.

**TESSELLATIONS.** See MIRRORS.

**THOMAS JEFFERSON'S WHEEL CIPHER MACHINE.** See CRYPTARITHMS, CRYPTOGRAPHY, CONCEALMENT CIPHERS, SUBSTITUTION CIPHERS, TRANSPOSITION CIPHERS, AND CODE MACHINES.

**THOMPSON'S LAMP.** See ZENO'S PARADOXES.

**TICTACTOE,** also called Noughts and Crosses, is a simple two-person game.

A board is set up as follows, and each player in turn places his mark (a *x* or an *o*) in one of the spaces. The goal is to get three marks in a row, horizontally, vertically, or diagonally:

If both players play rationally, the game can only end in a draw.

A traditional form, popular in China, Greece, and Rome, allows each player three markers. These markers are then placed on the board one at a time, as marks would be made in a standard game. However, after the three markers are placed on the board by each player, players take turns moving their markers about the board in an attempt to get three in a row. Only moves along adjacent orthogonals are permitted. Variations allow diagonal moves and/or two-space moves. Other variations expand the board to a 4 × 4 or 5 × 5 square and increase the number in a row needed to win to four or five.

An inversion, called toetacktick by Mike Shodell, gives the victory to the player who does not get three in a row.

Three-dimensional tictactoe, played on 3 × 3 × 3 or 4 × 4 × 4 cubes and allowing wins on the four main diagonals of the cube in addition to the standard rows, produces a game that cannot end in a draw for the 3 × 3 × 3 cube. Variations have allowed wins on crossing (intersecting) rows.

For similar play refer to Checkers; Geometric Problems and Puzzles.

BIBLIOGRAPHY

Gardner, Martin. The "Scientific American" Book of Mathematical Puzzles and Diversions. New York: Simon & Schuster, 1959.
Schuh, Fred. The Master Book of Mathematical Recreations. Trans. by F. Göbel; trans. and ed. by T. H. O'Beirne. New York: Dover, 1968.

**TIDDLY-WINK.** See DOMINOES.

**TIME PROBLEMS.** See ARITHMETIC AND ALGEBRAIC PROBLEMS AND PUZZLES.

**TOETACKTICK.** See TICTACTOE.

**TOPOLOGY** is the study of the properties of geometric figures that persist even when the figures are so distorted that all their metric and projective properties are lost, or the study of properties that remain invariant under continuous deformation.

A topological transformation of one geometrical figure A into another A' is given by the correspondence p ↔ p' between the points (p) of A and the points (p') of A' which has the following two properties: First, the correspondence is biunique (each point in A has only one corresponding point in A' and vice versa). Second, the correspondence is continuous in both directions (if any two points (p, q) of A are moved so that the distance between p and q approaches zero, then the distance between p', q' in A' will also approach zero, and vice versa).

Any property for a geometrical figure A that also holds for every figure into which A may be transformed by a topological transformation is called a topological property of A. Topology is the branch of geometry that deals only with the topological properties of figures.

Topology is a fairly recent development in geometry that began in the nineteenth century. A. F. Möbius (1790–1868) submitted a memoir on "one-sided" surfaces to the Paris Academy in 1858, but it was largely overlooked. Independently, J. B. Listing (1808–1882) made similar discoveries, and, at the urging of Karl Friedrich Gauss, published *Vorstudien zur Topologie* in 1847. These studies and the general atmosphere of the time set the groundwork for Bernhard Riemann (1826–1866) to come up with his theory of functions, which in turn stimulated the development of topology.

Henri Poincaré and other early innovators in the field were forced to rely on intuition. However, a surge of activity by L.E.J. Brouwer, O. Veblen, J. W. Alexander, and S. Lefschetz has brought topology within the framework of rigorous mathematics and connected it with the whole of mathematics.

Perhaps the easiest examples of topological transformations can be seen in terms of the deformations that occur when a piece of rubber is stretched. A geometric figure drawn on that piece of rubber would take on many different shapes. Yet it would remain a legitimate topological image or transformation as long as two points are not brought into coincidence and the piece of rubber is not ripped. It is possible to cut a figure during a topological transformation and sew it up again to produce a topological equivalent.

The following two figures are not topologically equivalent:

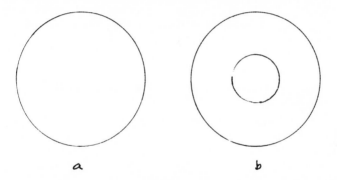

In *a* any closed curve can be deformed into a single point (the domain is simply connected). In *b* a closed curve around the center hole cannot be shrunk down into a single point (the domain is multiply connected). What is needed to make the two surfaces equivalent is to cut *b* so as to eliminate the center hole:

The Jordan curve is another important aspect of topological surfaces. It goes as follows: any simple closed curve in a plane divides the plane into exactly two domains, an inside and an outside:

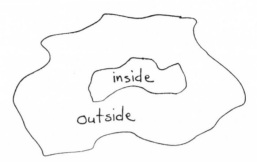

The *map-coloring problem* is one of the more famous problems in recreational topology. The question is whether or not four colors are both necessary and sufficient in order to color any map so that no two regions sharing a common border will have the same color. August Ferdinand Möbius (1840), Augustus de Morgan (1850), and Arthur Cayley (1878) all pointed out the difficulty of proving this one way or the other. In 1879, Alfred Bray Kempe published what he thought was a proof of it, but in 1890, Percy John Heawood found an error in the proof. It was not until 1976 that a proof that four colors were both necessary and sufficient was published.

Martin Gardner includes the following two topological puzzles in *Mathematical Magical Show*: First, what is the largest cube that can be completely covered on all six sides by folding around it a pattern cut from a three-inch-square sheet of paper (the paper must be of one piece; no overlapping is allowed)? The solution:

Second, how can two of the four matches forming the cocktail glass around the cherry be moved so that the glass is reformed in a different position and the cherry is outside the glass?

The solution:

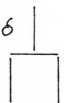

Gardner discusses the following more complex trick in *The Second "Scientific American" Book of Mathematical Puzzles and Diversions* (according to Gardner, it originated with Stewart Judah, a Cincinnati magician): A shoelace is wrapped securely around a pencil and a soda straw. When the ends of the shoelace are pulled, it appears to penetrate the pencil and cut the straw in half.

It works as follows: The soda straw is pressed flat and one end of it is attached with a rubber band to the end of an unsharpened pencil:

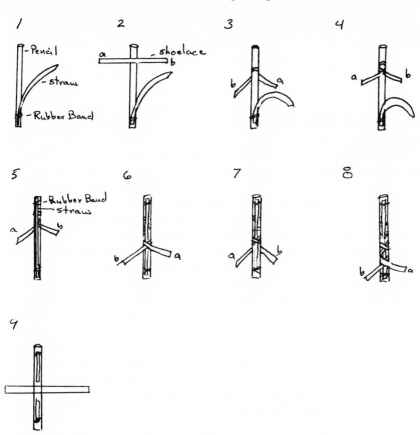

The straw is bent down from the tip of the pencil at a 45 degree angle. The middle of the shoelace is placed over the pencil and crossed behind the pencil. Whenever a crossing occurs in the winding, the same end is always overlapping the other end.

The ends are brought forward and crossed in front of the pencil. The straw is bent upward, so that it lies along the pencil, and is fastened to the top of the pencil with another rubber band. The shoelace is crossed above the straw, keeping end *a* on top of *b*. The two ends are wound behind the pencil for another crossing and then forward for a final crossing in front.

A spectator is asked to grip the pencil tightly, and the ends are given a quick pull.

This is what happens: Since the ends of the shoelace spiral around the pencil in mirror-image helices, the closed curve represented by the performer and the lace is not linked with the closed curve formed by the spectator and the pencil. The lace cuts the straw that holds the helices in place; then the helices annihilate each other.

Sprouts is a topological game originated by John Horton Conway and Michael Steward Paterson and discussed by Martin Gardner, *Mathematical Carnival*. *N* number of spots are put on a piece of paper (a small number, as even three leads to a complicated game). On each move a line is drawn joining two spots or one to itself, and a new spot is placed anywhere on the line. The line may have any shape, but cannot cross itself, a previously drawn line, or a previously made spot. No spot may have more than three lines emanating from it. The winner is the final player able to form a legitimate play (or, in the reverse, the first player unable to play).

Conway has offered proof that every game must end in at most $3n - 1$ moves, since each spot has at most three lives and each move kills two lives and adds one.

Conway also suggests a second (fixed) form called Brussels Sprouts, which uses crosses instead of spots. All the rules are the same, except that the lines are drawn from the ends of the crosses. Since every game must end in $5n - 2$ moves, if the standard game starts with an odd number of crosses, the first player always wins; the second player wins if it starts with an even number.

Martin Gardner explained the game in his *Scientific American* column and received numerous variations from his readers. Ralph J. Ryan III suggested replacing the spots with arrows extending from one side of a line and allowing new lines to be drawn only to the arrows' points. Gilbert W. Kessler combined spots and crossbars in a game called Succotash. Eric L. Gans replaced the spots with stars (*n* crossbars crossing at the same point). And Vladimir Ygnetovich offered the variation of allowing players a choice of whether to add one, two, or no spots to their line on each turn. (Refer to Nim for similar recreations.)

Here is a sample game of Sprouts:

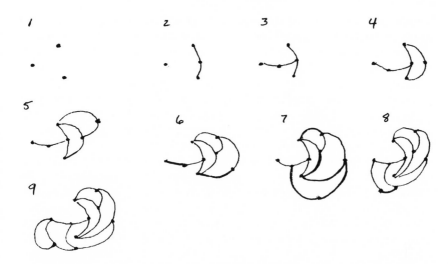

For similar recreations refer to Geometric Problems and Puzzles; The Möbius Strip.

BIBLIOGRAPHY

Courant, Richard, and Herbert Robbins. *What Is Mathematics? An Elementary Approach to Ideas and Methods.* New York: Oxford University Press, 1960.

Dudeney, Henry Ernest. *Amusements in Mathematics.* 1917. Reprint. New York: Dover, 1958.

Gardner, Martin. *Mathematical Carnival.* New York: Alfred A. Knopf, 1965.

_____. "Mathematical Games: The Topology of Knots, Plus the Results of Douglas Hofstadter's Luring Littery." *Scientific American* 249, no. 3 (Sept. 1983): 18–28.

_____. *Mathematical Magic Show.* New York: Alfred A. Knopf, 1977.

_____. *Mathematical Puzzles.* New York: Thomas Y. Crowell, 1961.

_____. *The Second "Scientific American" Book of Mathematical Puzzles and Diversions.* New York: Simon & Schuster, 1961.

Mendelson, Bert. *Introduction to Topology.* Boston: Allyn and Bacon, 1968.

Schaaf, W. L. "Number Games and Other Mathematical Recreations." In *Macropaedia: Encyclopaedia Britannica,* 1985.

**TOWER OF BRAHMA.** See THE TOWER OF HANOI.

**THE TOWER OF HANOI** is a manipulative puzzle involving Mersenne numbers.

It consists of three pegs fastened to a horizontal board, and eight circular disks, each a different size, with a hole in the middle. The disks are placed on one of the pegs, the disk with the largest radius on the bottom, each disk decreasing in radius to the smallest on the top. The object is to transfer disks from one peg to another, one at a time, so that no disk ever rests on a disk smaller than itself until the tower has been rebuilt on another peg.

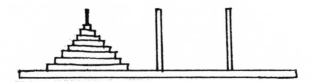

The puzzle was invented by the French mathematician Edouard Lucas and sold as a toy in 1883. It was originally sold under the name "Prof. Claus" of the College of "Li-Sou-Stian," an anagram for Prof. Lucas of the College of Saint Louis. He gave the following story its origin:

The Mandarin Claus, who came from Siam, claimed that during his travels in connection with the publication of Fer-Fer-Tam-Tam he encountered the great temple at Benares, which contained, beneath the dome of the center of the earth, a brass plate in which are fixed three diamond needles, each a cubit high and as thick as the body of a bee. On one of these needles God placed at the Creation sixty-four disks of pure gold, the largest disk resting on the brass plate, the others getting smaller and smaller up to the top. This is the Tower of Brahma. According to the laws of Brahma, which require that that the priests on duty must not move more than one disk at a time, and that no disk may be placed on a needle which already holds a smaller disk, the priests of the Tower of Brahma day and night unceasingly transfer the disks from one diamond needle to another. When the sixty-four disks shall have been thus transferred from the needle on which at the Creation God placed them to one of the other needles, then towers, temple, and priests alike will crumble into dust, and with a might thunderclap the world will vanish.

If the number of transfers the priests make in one second is 1, then they must work for $2^{64} - 1 = 18,446,744,073,709,551,615$ seconds, or more than 500,000 million years.

Eugene P. Northrop, *Riddles in Mathematics: A Book of Paradoxes,* points out that the number $2^{64} - 1$ has also been connected to the origin of chess. Apparently legend has it that an ancient Shah of Persia was so impressed with chess that he commanded its inventor to request any reward he wished. The inventor responded that he wished only to have one grain of wheat for the first square on the chessboard, two grains of wheat for the second, four for the third, eight for the fourth, and so on, or $1 + 2 + 2^2 + 2^3 + 2^4 + \ldots 2^{63} = 2^{64} - 1$ grains of wheat. The Shah, who apparently was not much of a mathematician, thought this a poor reward, at least until his advisors worked it out.

If two chessboards are placed together and this process is continued, the number obtained, minus one grain of sand, will be the largest known prime number: $(2^{127} - 1)$ or $170,141,183,460,231,731,687,303,715,884,105,727.$

It is not hard to prove that there is a solution, regardless of how many disks are on the tower, and that the minimum number of moves required can be determined by Mersenne numbers (numbers of the form $2^n - 1$), $n$ equalling the number of disks. A Mersenne number may be prime (having no factor except itself or 1) or composite (made up of two or more prime factors). A necessary though not sufficient condition that $2^n - 1$ be a prime is that $n$ be a prime. Therefore, all even perfect numbers have the form $2^{n-1}(2^n - 1)$ where both $n$ and $2^n - 1$ are prime numbers. As of 1974, only twenty-four perfect numbers were known. It is known that for every Mersenne prime there is a corresponding even perfect number and vice versa. However, it is not known if there are any odd perfect numbers or whether there is an infinity of perfect numbers.

Perfect numbers have many curious properties. For example, all perfect numbers are triangular. The sum of the reciprocals of the divisors of a perfect number, including the reciprocal of the number itself, always equals 2, e.g., for 6: $\frac{1}{1} + \frac{1}{2} + \frac{1}{3} + \frac{1}{6} = 2$.

Martin Gardner, *The "Scientific American" Book of Mathematical Puzzles and Diversions,* explains the discovery of D. W. Crowe of the University of British Columbia that the Tower of Hanoi and the Icosian Game are closely related. The Icosian Game was invented in the 1850s by Sir William Rowan Hamilton to illustrate a type of calculus that was similar to his theory of quaternions. Hamilton used this calculus (called the Icosian calculus) to solve a number of path-tracing problems on the surfaces of the five Platonic solids. The game was played on the edges of a dodecahedron. Hamilton sold the game to a dealer in London in 1859 for 25 pounds.

The basic game goes as follows: Start at any corner on the solid. Make a complete trip around the solid, visiting each vertex once and only once, and return to the starting point.

To demonstrate how the Tower of Hanoi is similar to the Icosian Game, start with a simple version of the Tower of Hanoi using only three disks, labeled, from top to bottom, A, B, C. The puzzle is solved by moving the disks in the following order: ABACABA. If the three coordinates of a cube (the edges extending horizontally, the edges extending vertically, and the edges extending in depth) are also labeled ABC, the path will have the same ABACABA order:

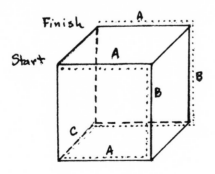

For similar games and puzzles refer to Chess Problems; Chinese Rings; The Fifteen Puzzle; Nim.

BIBLIOGRAPHY

Ball, W. W. Rouse. *Mathematical Recreations and Essays*. 1892. Rev. by H.S.M. Coxeter. London: Macmillan and Co., 1939.

————. *A Short Account of the History of Mathematics*. 1888. Reprint. London: Macmillan and Co., 1935.

Gardner, Martin. *The "Scientific American" Book of Mathematical Puzzles and Diversions*. New York: Simon & Schuster, 1959.

Kraitchik, Maurice. *Mathematical Recreations*. 2d rev. ed. New York: Dover, 1942.

Northrop, Eugene P. *Riddles in Mathematics: A Book of Paradoxes*. New York: D. Van Nostrand Co., 1944.

Schaaf, W. L. "Number Games and Other Mathematical Recreations." In *Macropaedia: Encyclopaedia Britannica*, 1985.

Steinhaus, H. *Mathematical Snapshots*. 3d ed. New York: Oxford University Press, 1969.

Tietze, Heinrich. *Famous Problems of Mathematics: Solved and Unsolved Mathematical Problems from Antiquity to Modern Times*. Trans. by Beatrice Kevitt Hofstadter and Horace Komm. New York: Graylock Press, 1965.

**TRACE PUZZLES.** See GEOMETRIC PROBLEMS AND PUZZLES.

**TRANSPOSITION CIPHERS.** See CRYPTARITHMS, CRYPTOGRAPHY, CONCEALMENT CIPHERS, SUBSTITUTION CIPHERS, AND CODE MACHINES.

**TRIANGULAR NUMBERS.** See NUMBER PATTERNS, TRICKS, AND CURIOSITIES.

**TRIBONACCI NUMBERS.** See NUMBER PATTERNS, TRICKS, AND CURIOSITIES.

**TRIPLICATION PROBLEM.** See DOMINOES.

**THE TRISECTION OF AN ANGLE** stands along with the duplication of a cube and the squaring of a circle as one of the three great construction problems of ancient geometry.

The challenge is simple. Find a method for trisecting any angle using nothing more than a straight-edge and compasses. The would-be solver is allowed to draw a straight line of indefinite length through two given distinct points and to construct a circle with a center at a given point and passing through a second given point. The equipment is simple, and the rules are minimal but rigid.

At first glance, the problem would seem to be easily solved. After all, similar problems involving the same rules of construction (the rules are said to have come from Plato), e.g., the bisection of an angle and the division of a line segment into any desired number of equal parts, are so simple as to be almost insulting.

At least as far back as the fifth century B.C. the Greek geometers set themselves to find the solution. They knew that certain angles could be easily trisected. Trisection of the right angle, for instance, is easily accomplished:

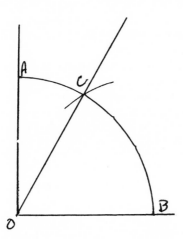

Draw arc *AB* and then, without altering the compass opening, place the compass point at *B* and draw an arc that intersects the other arc at *C*. The line from *O* through *C* trisects the right angle.

However, the Greek mathematicians could not find a method applicable to *any* angle.

In fact, it was not until 1837 that P. L. Wantzel (1814–1848), a French mathematician, published a complete, rigorous proof that a general method for trisecting any angle was *not* possible.

The proof comes not from playing with a straight-edge and compasses but from algebra. Simplified explanations of the proof are contained in *Famous Problems of Elementary Geometry* by Felix Klein; *The Trisection Problem* by Robert C. Yates; and "Mathematical Games: The Persistence (and Futility) of Efforts to Trisect the Angle" by Martin Gardner.

Briefly, the proof is based on the fact that elementary geometry has no general method of algebraic analysis. Thus, in analyzing what geometric constructions are theoretically possible, it is necessary to consider rational operations (addition, subtraction, multiplication, and division), all of which are possible upon two given segments through the use of proportions, so long as an auxiliary unit-segment is introduced for multiplication and division; and irrational operations (algebraic and transcendental), which can only be constructed if they involve square roots with a limited number of operations. Thus, if an equation corresponding to a given construction is not solvable by a finite number of square roots, that construction is not possible with a straight-edge and compasses.

In other words, since straight lines on the Cartesian plane are graphs of linear equations, and circles are graphs of quadratic equations, there are five and only five operations that can be performed on line segments using only compasses and a straight-edge: addition, subtraction, multiplication, division, and extraction of square roots. Thus, for example, any angle equal to $360/n$ degrees where $n$ can be divided by 3 cannot be trisected.

Nevertheless, the seeming simplicity of the problem continues to lure would-be solvers. Robert C. Yates, *The Trisection Problem*, discusses five of the more interesting ones since 1900 and lists twelve others since Wantzel's proof. One of the more famous ones (also discussed in Gardner's "Mathematical Games"), was put forth in 1931 by Reverend Jeremiah Joseph Callahan, president of Duquesne University in Pittsburgh. It went as follows:

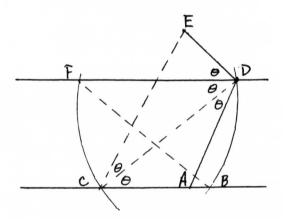

Lines *BC* and *DF* are drawn parallel to one another. With any point, e.g., *D*, as center, describe the circular arc *FC*. Then, with the same radius and center *F*, draw arc *DB*. Construct the angle *DCE* equal to angle *DCA*. Draw *DA* parallel to *EC* and *DE* parallel to *FB*. Then, *DF* and *DC* trisect angle *ADE*.

What is the flaw? Well, instead of trisecting an angle, Callahan has drawn its triple, e.g., *ADE* is the triple of *DCA*. However, because of Callahan's highly respected position in education, he received a great deal of publicity. *Time*, for instance, ran his picture and a favorable account of his discovery.

From the beginning, intelligent mathematicians attempted to solve the problem through the use of more complex curves than the circle (thus, of course, breaking the rules).

Hippias of Elis (c. 420 B.C.) came up with a curve called the quadratrix, which allows the trisection of an angle. It works as follows:

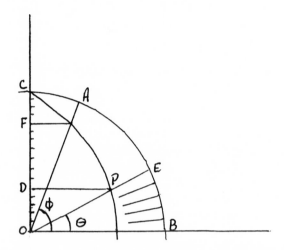

$COB$ is a quadrant of a unit circle. Bring the point $D$ along the line from $O$ to $C$ at a constant rate. At the same time, bring point $E$ along the arc from $B$ to $C$. Then, the horizontal line through $D$ meets $OE$ at $P$, and the path described by $P$ is the quadratrix. Thus, the ratio of the lengths of any two arcs $BE$ and $BA$ is the same as the ratio of their corresponding segments on $OC$, i.e., $OD/OF = BE/BA = \theta/\phi$. Therefore, once the curve is drawn, if $AOB = \phi$ is the angle to be trisected, all that is needed is to take $OD = (\frac{1}{3})OF$, and $(\frac{1}{3})(OF/OF) = \theta/\phi$, or $\theta = \phi/3$.

Nicomedes (c. 200 B.C.) came up with a curve called the conchoid, which offers a simple solution to the trisection problem. It works as follows:

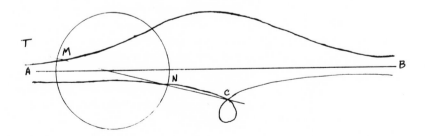

Draw line *AB*. A circle moves along this line with its center always on the line. Through a fixed point *C*, not on the line, draw line *CT*. The paths created by points *M* and *N* as the circle passes along the line form the conchoid.

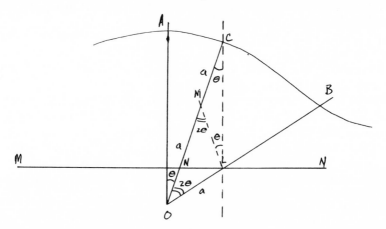

In the above illustration, ∡ *AOB* is trisected by placing the angle with vertex at *O* and drawing perpendicular line *MN* cutting *OB* at *L* so that *OL* = *a*, the projection value: *cos(AOB)*. Then, using *2a* as the radius of the generating circle, draw the conchoid *ACB*. Construct the parallel to *OA* at *L*, which will meet the curve at *C*. *OC* then trisects *AOB*.

The proof of this is as follows: ∡ *AOC* = *LCO* = θ. Since *CN* = *2a* (definition of the curve) and *CLN* is a right angle, then the segment from the midpoint *M* of *CN* to the right angled vertex *L* is length *a*. Therefore, triangles *CML* and *MLO* are both isosceles. Thus, ∡ *AOC* = ∡ *OCL* = ∡ *MLC* = θ. However, ∡ *OML* = 2θ, because it is the exterior angle of △ *CML*. And therefore, ∡ *MOL* = 2θ, and ∡ *AOB* is trisected by *OC*.

Pappus (c. A.D. 300) included the use of the hyperbola among his solutions to the problem. It works as follows:

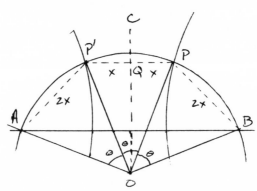

Form a unit circle with its center at *AOB* and bisect it with *OC*. Allow point *P* to move so that its distance is always twice as far from *B* as from *OC*. This will then cause *P* to trace out a branch of a hyperbola with the line *OC* as the directrix and the point *B* as the focus. This branch is then reflected in *OC* so that *P'* corresponds to *P*.

The points of intersection of the hyperbola and the unit circle are trisecting points of arc *AP'PB*, because if $PQ = x = P'Q$, then $PB = P'A = 2x$ and the three isosceles triangles *AOP'*, *P'OP*, and *POB* are congruent to each other with equal angles at their common vertex *O*.

The rectangular equation of the curve is derived as follows: Let *AB* and *OC* stand for the *X* and *Y* axes. Let *2c* denote the distance *AB*, thus giving *B* the coordinates *(c,O)*. Then, to state that *PB* must always be twice the distance *PQ*, give *P* coordinates *(x,y)*, and derive the following formula: $\sqrt{(x-c)^2 + y^2} = 2x$ or $y^2 - 3x^2 - 2cx + c^2 = 0$.

By simultaneously solving this equation and that of the circle, the location of the trisecting point *P* can be determined.

Blaise Pascal (1623–1662) came up with the limaçon, another curve with trisecting possibilities. René Descartes (1596–1650), founder of modern analytic geometry, attempted to solve the trisection problem with the following equation: $x^3 - 3x - 2a \neq 0$. This, he said, can be represented as the *x*-coordinates of the points of intersection of a parabola and a circle. Isaac Newton (1642–1727), *Geometria*, bk. III, ed. Schooten, Amsterdam, 1659, p. 91 (included in W. W. Rouse Ball, *Mathematical Recreations and Essays*), also employed the hyperbola, as did Alexis Claude Clairaut (1713–1765).

Along with these curves came devices for drawing them. The graduated ruler is perhaps the simplest. Others range from the compasses of H. Hermes (1883) and the three-bar apparatus useful in drawing Pascal's limaçon to the complex pantograph of Ceva and Lagarrique.

There are, of course, many ways to trisect an angle approximately. Perhaps the simplest is by Hugo Steinhaus, *Mathematical Snapshots*. First bisect the angle. Then divide a chord of a half-angle into three equal parts. The radius cutting two-thirds off the chord is a close approximation of the trisection of the original angle.

For similar mathematical play refer to The Duplication of the Cube; Geometric Problems and Puzzles; Squaring the Circle.

## BIBLIOGRAPHY

Ball, W. W. Rouse. *Mathematical Recreations and Essays*. 1892. Rev. by H.S.M. Coxeter. London: Macmillan and Co., 1939.

————. *A Short Account of the History of Mathematics*. London: Macmillan and Co., 1888.

Behnke, H., F. Bachmann, K. Fladt, and H. Kunle. *Fundamentals of Mathematics. Vol. 2: Geometry*. Cambridge, Mass.: MIT Press, 1974.

Bell, E. T. *The Development of Mathematics*. New York: McGraw-Hill, 1945.

Gardner, Martin. "Mathematical Games: The Persistence (and Futility) of Efforts to Trisect the Angle." *Scientific American* 214 (June 1966): 116–120.

Graef, Edward V., and V. C. Harris. "On the Solutions of Three Ancient Problems." *Mathematics Magazine* 42 (Jan. 1969): 28–32.

Klein, Felix. *Vorträge Über Ausgewählte Fragen Der Element Argeometrie Ausgearbeitet Von F. Tägert*. 1895. Reprint. *Famous Problems of Elementary Geometry: The Duplication of the Cube, the Trisection of an Angle, the Quadrature of the Circle*. Trans. by Wooster Woodruff Beman and David Eugene Smith. 2d ed. Rev. by Raymond Clare Archibald. New York: Chelsea Publishing Company, 1962.

Steinhaus, H. *Mathematical Snapshots*. New York: Oxford University Press, 1969.

Tietze, Heinrich. *Gelöste und ungelöste mathematische Probleme aus alter und neuer Zeit*. Munchen: C. H. Beck'sche Verlagsbuchhandlung, 1959. Trans. as *Famous Problems of Mathematics: Solved and Unsolved Mathematical Problems from Antiquity to Modern Times*. Trans. by Beatrice Kevitt Hofstadter and Horace Komm. New York: Graylock Press, 1965.

Yates, Robert C. *The Trisection Problem*. National Council of Teachers of Mathematics, 1971.

**TRUTH AND LIES PROBLEMS.** See LOGICAL PROBLEMS AND PUZZLES.

**TWISTED PATH CIPHER.** See CRYPTARITHMS, CRYPTOGRAPHY, CONCEALMENT CIPHERS, SUBSTITUTION CIPHERS, TRANSPOSITION CIPHERS, AND CODE MACHINES.

**TYPEWRITER CODES.** See CRYPTARITHMS, CRYPTOGRAPHY, CONCEALMENT CIPHERS, SUBSTITUTION CIPHERS, TRANSPOSITION CIPHERS, AND CODE MACHINES.

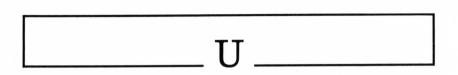

# U

**UTILITIES PUZZLE.** See GRAPHS AND NETWORKS.

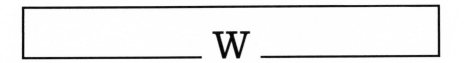

**W**

**WORLD CALENDAR.** See CALENDARS.

**WYTHOFF'S GAME.** See NIM.

# Z

**ZENO'S PARADOXES,** also called the paradoxes of Zeno, the sophisms of Zeno, and Zeno's sophisms, are a number of arguments designed to point out a paradox central to the science of geometry.

About 2,500 years ago, Zeno of Elea, a countryman and disciple of Parmenides, born about 490 B.C., put forth a number of problems dealing with space and time. The four most important are the Dichotomy (or Racecourse), Achilles and the Tortoise, the Arrow, and the Stadium (or Moving Rows). These problems point out that certain assumptions held by Zeno's countrymen result in contradictory conclusions, i.e., that either the Pythagorean conception of unit or the Anaxagorean conception of infinite divisibility must be abandoned, because the theory of units and infinite divisibility cannot both be accepted.

Strictly speaking, Zeno's paradoxes are sophisms, i.e., fallacies in which an error has been knowingly committed, and are, in fact, a form of dialectic. According to Bertrand Russell, *Wisdom of the West,* Zeno was the first to use dialectic argument systematically. What Zeno proves with his paradoxes (sophisms) is that the conclusions he derives from his (Pythagorean) assumptions are not only untrue but impossible, and thus the assumptions from which the conclusions follow are impossible.

In order to understand the importance of Zeno's paradoxes, it is necessary to place them in the context of the history of philosophy, beginning with some of the theories of Pythagoras.

Pythagoras settled in Croton, in Southern Italy, around 530 B.C., and remained there until 510 B.C. At that time, following a revolt against his school, he retired to Metapontion. Among other concepts of mathematics, he discovered numerical relations between musical intervals which correspond to harmonic progression (e.g., $2 : \frac{4}{3} : 1$). This discovery may well have been what led him to the idea that all things can be reduced to numbers, which, in turn, can be represented by arrangements of stones or pebbles (units). Furthermore, Pythagoras hypothesized, the air keeps the units distinct, and the units are what give measure to the air. (For more discussion of Pythagoras' reduction of all reality to numbers refer to Numerology.)

Heraclitus of Ephesus (sixth century B.C.), then, basing his views on the philosophy of Pythagoras and others, came up with the idea that the real world consists of a balanced adjustment of opposing tendencies, i.e., behind the apparent strife between opposites, according to an underlying mathematics, lies a hidden harmony. And this harmony is one of flux—all things are involved in this motion. As with all of the other theories previous to Parmenides, Heraclitus' view explains reality as material forms separated by empty space.

Parmenides, a native of Elea in Southern Italy, founder of the Eleatic school of philosophy, was the first to question this assumption. Parmenides' main thrust was that the claim that all things are made up of some basic stuff (it is) and at the same time that there is such a thing as empty space (it is not) is a contradiction. In other words, Parmenides said that previous philosophers were making a mistake in saying that "what is not is," i.e., that empty space (nothing) exists. According to Parmenides, nothing (empty space) cannot exist. Furthermore, what cannot be cannot be thought of, since, according to Parmenides, one cannot think of nothing. To solve this problem, Parmenides suggests a world where there is no empty space or even a space of lesser density. Rather, he postulates, the world is solid, finite, uniform, without time, motion, or change. This view of Parmenides, then, is the opposite of the view of Heraclitus (one claiming empty space and constant motion, the other denying empty space and motion).

In attempting to find a compromise between Parmenides and Heraclitus, Empedocles (first half of the fifth century B.C.) suggested the idea of "root" substances (these were later called by Aristotle "elements": water, air, fire, and earth), and something to cause these substances to mix (love and strife).

Anaxagoras then put forth the idea that each of the basic elements exists in every unit of material, and defended this idea with the concept of "infinite divisibility," which in turn allowed him to claim that no matter how much a unit is divided it will still contain all of the basic elements.

At this time Zeno came forth with his four paradoxes of motion to show that, if the Eleatic doctrine of Parmenides is unsatisfactory, other theories lead to even stronger objections.

In Achilles and the Tortoise, Zeno set up a hypothetical handicap race between Achilles and a tortoise. The premise is that in a handicap race Achilles can never catch up with the tortoise. If the tortoise starts at some distance down the track, then by the time Achilles runs up to where the tortoise started the tortoise will have moved some distance ahead.

In other words, if space is made up of individual units that are infinitely divisible, then the pursuer must pass through each point the pursued passes through, and while the pursuer is passing through these points, the pursued will be moving ahead, creating new points to be passed through. In this sense, motion, once started, cannot stop.

In the Dichotomy (Racecourse), Zeno asserted that motion cannot ever start, because one can never traverse an infinite number of points in a finite amount of time.

A          D          C                              B

To go from point A to point B it is first necessary to go to point C; to reach point C it is first necessary to reach point D; and so on. If one assumes that space is infinitely divisible, a finite length contains an infinite number of points. Before one can travel from one point to another, it is always necessary to travel to a point between the two. Therefore it is impossible ever to reach the second point, or, in other words, ever to start motion.

These two paradoxes, then, disprove the theory of a line consisting of an infinite number of units.

In the Stadium (Moving Rows), Zeno starts with three equal segments of lines (each made up of the same finite number of units). One line is at rest. The other two are moving past each other at equal rates of speed. The three lines all come together as the two moving lines pass the stationary line. The relative velocity of each of the moving lines to each other is twice as great as the relative velocity of each of the moving lines with the stationary line. If time is made up of units as well as space, then speed is measured by the number of points that move past a given point in a given number of moments. Thus, one line passes half the length of the stationary line at the same time that it passes the entire length of the moving line. Therefore the latter time is twice the former time. However, to reach their position alongside of each other, the two lines take the same amount of time. Therefore moving lines move twice as fast as they move.

In the Arrow, Zeno postulated that an arrow in flight is always at rest. What Zeno meant by this is that at any moment in time, if we grant indivisible units of time, the arrow occupies a space equal to itself and so does not move. In this case, motion cannot even start.

These last two paradoxes, then, show that positing a finite number of units in a line results in contradictions as great as positing an infinite number of units in a line.

In the process of pointing out the flaws in the Pythagorean theories of unit, Zeno raised questions about time, space, and motion that even today remain enigmas, though Georg Cantor (1845–1918) offered a solution accepted by many contemporary mathematicians (*see* Paradoxes of the Infinite).

Martin Gardner, *Martin Gardner's Sixth Book of Mathematical Games from*

*"Scientific American,"* gives a simple explanation of how modern mathematicians would solve the paradox of Achilles and the Tortoise (a problem involving a converging series). The sum (limit) of a converging series is the number that the value of the series *approaches* as the number of its terms increases without limit, e.g., the sum of a halving series ($\frac{1}{2} + \frac{1}{4} + \frac{1}{8} + \frac{1}{16}$ . . .) is *1*. In some cases the value of an infinite series goes beyond its limit, e.g., the sum of an alternating halving series ($\frac{1}{2} - \frac{1}{4} + \frac{1}{8} - \frac{1}{16}$ . . .) is $\frac{1}{3}$. The point is that every infinite series that comes to an end will always have a partial sum that differs ever so slightly from the limit.

Gardner offers a simple solution for a converging series when the terms decrease in geometric progression. Let $x$ equal the entire series. Since in a halving series each term is twice as large as the next, multiply each side of the equation by 2: $2x = 2(\frac{1}{2} + \frac{1}{4} + \frac{1}{8} + \frac{1}{16} + \ldots)$ or $2x = 1 + \frac{1}{2} + \frac{1}{4} + \frac{1}{8} + \frac{1}{16} + \ldots$, since the new series beyond 1 is the same as the original series, $2x = 1 + x$, which reduces to $x = 1$.

If we assume an arbitrary geometric progression in the race between Achilles and the tortoise (say that Achilles runs ten times as fast as the tortoise and the tortoise has a 100-yard lead), and that they move at a uniform speed (none of these assumptions in any way negates the intent of Zeno in posing the problem), then after Achilles has gone 100 yards the tortoise will have gone 10, and so on. By applying the theory of limit to a converging series, we come up with *100 + 10 + 1 + .1 + .01 + .001 + . . .*, which equals *111.111 . . .* or *111 and $\frac{1}{9}$* yards.

Martin Gardner, *Wheels, Life and Other Mathematical Amusements,* includes Austin's Dog, a variation on Zeno's paradox of Achilles and the Tortoise that was mailed to Gardner by A. K. Austin of the University of Sheffield, England. The problem is as follows:

A boy, a girl, and a dog all start from the same spot on a straight road. The boy walks forward at four miles per hour. The girl walks forward at three miles per hour. And the dog trots back and forth between them at ten miles per hour, reversing its direction instantaneously. After an hour has passed, where is the dog and which way is it facing?

Austin's answer is as follows: The dog may be at any point between the boy and girl and may be facing either direction. The reasoning is that, if all the motions are time reversed, the three will return at the same time to the starting point.

The original paradox was published in *Mathematics Magazine* 44, no. 1 (Jan.–Feb. 1971): 56, and received the following replies, published in *Mathematics Magazine* 44, No. 4 (Sept.–Oct. 1971): 238–239:

I. Comment by M. S. Klamkin, Ford Motor Company.

I disagree with the proposer's solution. While I agree that the motion is reversible from any initial starting position in which the three participants are not at the same location, it is not possible to start the motion when all three start from the same location. The dog would have a nervous breakdown attempting to carry out his program. If one is not convinced, let the initial starting distance between the boy and the girl be $\epsilon$ (arbitrarily small), then one can show that the number of times the dog reverses becomes arbitrarily large in a finite time.

An analogous situation occurs in the well known problem of the four bugs pursuing each other cyclically with the same constant speed and starting initially at the vertices of a square. At any point of their motion (except when together), the motion is reversible by reversing the velocities. However, when together, the directions of the velocities are indeterminate and thus cannot reverse without further instructions.

II. Comment by Leon Bankoff, Los Angeles, California.

The published solution is based on the fallacious *a priori* assumption that the dog can be placed at a point between the positions of the boy and the girl. The conditions of the problem preclude the possibility of this occurrence. At any moment after the simultaneous start, the arrangement of the three will be girl, boy, dog—never girl, dog, boy.

III. Comment by Charles W. Trigg, San Diego, California.

At the end of one hour the dog is 10 miles from the starting point and facing away from it, since it never is between the boy and girl. That is, unless its position on the circle around the earth on which the straight road lies is considered. In that event it is between the boy and girl, leaving the boy behind and catching up with the girl. It will not need to change its direction for a long time, not until it has traversed the circumference and continuing on, passed the girl and caught up with the boy. They all should be well exhausted by that time.

IV. Comment by Lyle E. Pursell, University of Missouri-Rolla.

Quickie Q503 is self-contradictory. If the boy, girl and dog start from the same point at speeds of *4, 3,* and *10,* respectively, then at any *positive* time, *t, no matter how small,* the dog will not be between the boy and the girl. Hence, the dog cannot run "back and forth between the two of them, going repeatedly from one to the other and back again" as the author prescribes.

The author's solution to the problem looks like a proposal to sum an infinite series by starting with the "last" term! Since, if later the three reverse their motion as the author suggests in his solution, then the dog must reverse his direction infinitely many times before the boy and the girl get back to the starting point.

Both Martin Gardner and Wesley C. Salmon defend Austin's solution, pointing out that when time-reversal takes place it is impossible to know where the dog finishes. Since the dog cannot be told how to start, it can start any way it wishes, and thus end at any point between the boy and the girl. This is exactly what Austin claims in his solution.

As both Gardner and Salmon point out, Austin's Dog is similar to the time-reversal problem about the bird that flies back and forth between two converging trains. The trains, starting at points *A* and *B,* 30 miles apart, converge on each other at, say, 15 miles per hour, until they collide at *C* (it takes one hour). A bird, starting at *A,* flies back and forth at 60 miles per hour between the trains until they collide. How long is the bird's path? Since the bird flies for an hour, it must be 60 miles.

However, if it is not established that the bird must be at a certain point in a reversal of the problem, i.e., when the trains move back to points $A$ and $B$, then the backward flying bird can only be located somewhere between the trains.

Such problems as these are often classified as "runner" problems. They raise the same types of questions that Zeno's runners raised.

Another group of problems (sophisms), categorized as "supertasks," the type of tasks performed by "infinity machines," has been raised by such people as James F. Thompson and Max Black. Thompson's Lamp goes as follows: The lamp is turned on for 1 minute, then off for $\frac{1}{2}$ minute, on for $\frac{1}{4}$ minute, and so on. At the end of 2 minutes, will the lamp be on or off? The question raised is this: If Achilles can run an infinity of halfway points in 2 minutes (and actually climb upon the back of the tortoise, as Lewis Carroll satirizes), why cannot Thompson's lamp be turned on and off an infinite number of times in 2 minutes, ending exactly at 2 minutes? However, if this is possible, then there is a final counting number—absurd.

Black's machine transfers a marble from tray $A$ to tray $B$ in 1 minute, then puts it back in tray $A$ in $\frac{1}{2}$ minute, then puts it back in tray $B$ in $\frac{1}{4}$ minute, and so on. At the end of 2 minutes, where is the marble? If it is in tray $A$, then an odd number of moves has taken place, if in $B$ an even number, in which case the final counting number must be either odd or even.

For more in-depth discussion of Zeno's paradoxes refer to Wesley C. Salmon, *Zeno's Paradoxes*. The paradoxes have raised such an unending debate that a final solution or elimination for them can hardly be said to be near (though many contemporary mathematicians feel that one or another approach has solved them). It should be pointed out that many prominent thinkers, including Bertrand Russell, question calling the paradoxes "sophisms."

For similar mathematical discussion refer to Paradoxes of the Infinite.

BIBLIOGRAPHY

Bakst, Aaron. *Mathematics: Its Magic and Mastery*. New York: D. Van Nostrand Co., 1941.
Collins, A. Frederick. *Fun with Figures*. New York: D. Appleton and Co., 1928.
Gardner, Martin. *Aha! Gotcha: Paradoxes to Puzzle and Delight*. San Francisco: W. H. Freeman and Co., 1982.
———. *Martin Gardner's Sixth Book of Mathematical Games from "Scientific American."* San Francisco: W. H. Freeman and Co., 1971.
———. *Wheels, Life and Other Mathematical Amusements*. San Francisco: W. H. Freeman and Co., 1983.
Grunbaum, Adolf. *Modern Science and Zeno's Paradoxes*. Middletown, Conn.: Wesleyan University Press, 1967.
Kline, Morris. *Mathematics in Western Culture*. New York: Oxford University Press, 1953.

Northrop, Eugene P. *Riddles in Mathematics: A Book of Paradoxes*. New York: D. Van Nostrand Co., 1944.

Russell, Bertrand. *Wisdom of the West*. London: Crescent Books, 1977.

Salmon, Wesley C. *Zeno's Paradoxes*. New York: Bobbs-Merrill Co., 1970.

Schaaf, W. L. "Number Games and Other Mathematical Recreations." In *Macropaedia: Encyclopaedia Britannica*, 1985.

Tietze, Heinrich. *Famous Problems of Mathematics: Solved and Unsolved Mathematical Problems from Antiquity to Modern Times*. Trans. by Beatrice Kevitt Hofstadter and Horace Komm. New York: Graylock Press, 1965.

**ZENO'S SOPHISMS.** See ZENO'S PARADOXES.

# BIBLIOGRAPHY

William L. Schaaf has been publishing a continually updated *Bibliography of Recreational Mathematics*. For further exploration of most of the entries in this book that is the place to begin. He has also written a survey of recreational mathematics titled "Number Games and Other Mathematical Recreations" for the 1985 *Encyclopaedia Britannica*.

Since each entry in this book has its own bibliography, only the more important general works are included below.

Ball, W. W. Rouse. *Mathematical Recreations and Essays*. 1892. Rev. by H.S.M. Coxeter. London: Macmillan and Co., 1939.

Dudeney, Henry Ernest. *Amusements in Mathematics*. 1917. Reprint. New York: Dover, 1958.

———. *The Canterbury Puzzles*. 1919. Reprint. New York: Dover, 1958.

———. *536 Puzzles and Curious Problems*. Reprint. Ed. by Martin Gardner. New York: Dover, 1967.

Gardner, Martin. *Aha! Gotcha: Paradoxes to Puzzle and Delight*. San Francisco: W. H. Freeman and Co., 1982.

———. *Martin Gardner's New Mathematical Diversions from "Scientific American."* New York: Simon & Schuster, 1966.

———. *Mathematical Carnival*. New York: Alfred A. Knopf, 1977.

———. *Mathematical Circus*. New York: Alfred A. Knopf, 1979.

———. *Mathematical Magic Show*. New York: Alfred A. Knopf, 1977.

———. *Mathematical Puzzles*. New York: Thomas Y. Crowell Co., 1961.

———. *Mathematics, Magic, and Mystery*. New York: Dover, 1956.

———. *The "Scientific American" Book of Mathematical Puzzles and Diversions*. New York: Simon & Schuster, 1956.

———. *The Second "Scientific American" Book of Mathematical Puzzles and Diversions*. New York: Simon & Schuster, 1961.

Loyd, Sam. *Sam Loyd's Cyclopedia of Puzzles*. Reprint. Ed. by Martin Gardner. *Mathematical Puzzles of Sam Loyd, Vol. I and Vol. II*. New York: Dover, 1959.

Kinnard, Clark, ed. *Encyclopedia of Puzzles and Pastimes*. New York: Citadel Press, 1946.

Kraitchik, Maurice. *Mathematical Recreations*. 2d rev. ed. New York: Dover, 1953.

Madachy, Joseph S. *Mathematics on Vacation*. New York: Charles Scribner's Sons, 1966.

Phillips, Herbert. *Problem Omnibus, Vol. I and Vol. II*. New York: Dover, 1962.

Schuh, Fred. *The Master Book of Mathematical Recreations*. Trans. by F. Göbel; trans. and ed. by T. H. O'Beirne. New York: Dover, 1968.

Steinhaus, H. *Mathematical Snapshots*. 3d Am. ed., rev. and enlarged. New York: Oxford University Press, 1969.

A number of magazines include recreational mathematics. Before listing some of the more important ones, Martin Gardner's column, "Mathematical Games," which appeared in *Scientific American* from the 1950s into the 1980s, deserves special mention. Some of the other important magazines include the following: *American Mathematical Monthly, Fibonacci Quarterly, Journal of Recreational Mathematics, Mathematical Gazette, Mathematics Magazine, Mathematics Teacher, National Mathematics Magazine, Recreational Mathematics Magazine, Scientific Monthly,* and *Scripta Mathematica*.

It should be stressed that the above bibliography, as well as those for each entry, is both subjective and incomplete, meant as a starting place for further research.

# INDEX

Page numbers in *italics* refer to main entries in the dictionary.

# ABOUT THE AUTHOR

HARRY EDWIN EISS is Assistant Professor of English at Eastern Michigan University, Ypsilanti, Michigan. He is the author of *Dictionary of Language Games, Puzzles, and Amusements* (Greenwood Press, 1986), has published articles on language arts instructions, and his poems have appeared in several anthologies.